GeoSystems Today: An Interactive Casebook

Robert E. Ford
Westminster College of Salt Lake City
Salt Lake City, Utah

James D. Hipple
University of Missouri
Columbia, Missouri

John Wiley & Sons, Inc.
New York • Chichester • Weinheim • Brisbane • Singapore • Toronto

ISBN 0-471-19598-7

Printed in the United States of America

10 9 8 7 6 5 4 3 2 1

Printed and bound by Courier Kendallville, Inc.

Acknowledgements

I acknowledge with gratitute the contributions of several people involved with this print and web-based project. First I thank my co-author James D. Hipple of the University of Missouri-Columbia who not only contributed an excellent case study on Urban Ecology but also did the time-consuming web-design and HTML coding of the online interactive materials. He also produced the layout and pre-press design for the print casebook. Without his patience and technical skills this project would not have occurred.

Five of the case studies were contributed by colleagues from both academic and community circles—without their assistance this project would not have achieved the breadth it has attained. I thank specifically George F. Hepner and Mark V. Finco, of the University of Utah, Department of Geography, who produced a wonderful case on the US-Mexico Border. Michael W. Hernandez was our specialist on the Wasatch Fault and W. Scott White enlightened us on the Mississippi/Missouri River Floods of 1993. Robert J. Moye and Sandra A. Zicus, are environmental educators and consultants based in Salt Lake City. Between the two of them they produced excellent case studies on Saline Lakes as well as the Petroleum Geology of the Overthrust Belt and Persian Gulf regions.

Finally, I must thank Clifford Mills and Fadia Stermasi, the Geoscience Editor and New Media Specialist for John Wiley & Sons, Inc., who supported us through the long gestation and production phase of this project. Several of their staff were helpful at crucial phases in finding graphics, design work and production; specific thanks go to Tammy McGinnis, Jill Hilycord, Marge Graham and others.

And of course, our gratitude and thanks goes to our families who often suffered through this project's development and final completion.

Robert E. Ford
Sandy, Utah
December 2, 1998

Table of Contents

GeoSystems Today: an Interactive Casebook presents eight case studies in the geosciences that illustrate the complex integrated nature of earth systems. Another key focus is on exploration of the **human dimensions of global change (HDGC)**. The printed case material is supplemented by extensive Internet based graphic and interactive learning resources that liven up the cases in ways text alone cannot do. The goal is not only to pique your curiosity but to actively engage you—the student—in structured inquiry of geoscience problems affecting real people, places and earth systems on our small Planet. We particularly want to raise questions that lead you to struggle with the complex interaction of positive and negative forces and variables that enhance or detract from achievement of that illusory state of human-environmental balance referred to as sustainable development.

We also hope that exploration of the case material, particularly the online resources, will expose you to exciting new "hi-tech" ways of doing science. We want you to experiment with and observe the tools, techniques, conceptual models and findings of contemporary **global environmental change (GEC)** and **Earth System Science (ESS)** researchers. In the cases you will observe how real geoscientists define, study and analyze complex interdisciplinary problems and issues. You will also be confronted by what differing individuals and organizations, of various persuasions and opinions, are saying and doing about their findings. This will expose you to the "messy" side of science as it is applied in policy-making and decision-making. The ultimate goal is to not only increase skills and knowledge but to heighten personal involvement and commitment to solving some of Planet Earth's great social and environmental problems.

Level of Difficulty and Pedagogical Design

The printed casebook starts at a very basic introductory level and assumes little previous exposure to the topic beyond what is available in introductory college textbooks. Furthermore, the print material and most of the online resources do not require quantitative skills. Some of the supplementary online resources offer suggestions for designing activities/exercises that demonstrate the value of "hi-tech" tools such as computer modeling, image analysis, geographical information systems, and remote sensing. Your instructor can use these materials to design more advanced analytical activities. An online Teacher's Integrator is provided to assist instructors and students. It includes curriculum planning assistance such as suggested prior textbook reading assignments, answers to activities and exercises, correlations with the National Science Education and Geography Standards and supplementary notes and references.

The cases were designed to supplement introductory college-level textbooks in regional/ physical geography and geology, environmental science, earth system science, and even anthropology or area studies. The cases could become "jump-off" points for individual or group projects and papers, serve as springboards for discussions, seminars, supplement lectures, lab activities, or become take-home assignments. The online graphics and photos can also enhance visual presentations by either students or instructors (under certain copyright use limitations). The slide shows and virtual tours provide a "virtual" field experience where onsite study is not possible. The virtual tours could also be used as background reading prior to a real field study.

The design strategy explicitly recognizes that learners and teachers work in a diversity of ways and settings. Some learn by doing, others are visual learners, while yet others excel at reading and library research or quantitative analysis. Materials are available for every type of learner and teacher. Furthermore, we recognize that many users face limitations in access to the Internet or lack high quality computing power. Consequently the printed casebook uses simple textual material alone; for many learning situations this will suffice. The supplementary online resources have been designed for slow modem access and easy printing and navigation. Finally, our hope is that this casebook project will become an expanding resource base that is easily accessible, flexible, academically rigorous, interactive, graphically appealing and constantly updated and improved by you—the user.

Organization of the Case Material

Each printed case (in the book) starts with a a list of **Questions to be Explored** followed by **Key Learning Outcomes**. An **Overview** and **Background** section then introduces the essential geosciences concepts and issues followed by a section entitled **The Human Dimension**. The latter section presents an integrated perspective on the "nature-society" linkages and policy implications of the case. At the end are **Some Recommended Readings** for more indepth study. After you study the basic written text, and possibly do some background reading in a related textbook such as *The Blue Planet: An Introduction to Earth System Science,* you should take an online **Virtual Tour** and do the **Self-Study Exercise** found at the end of each case. At this point you should be ready to discuss the **Conceptual Questions** at the end of each case.

A suggestion—though the conceptual questions could be discussed with only the textual material as background, exploration of the online material as well would deepen your understanding and capacity to contribute meaningfully to a discussion. You are also strongly encouraged to explore some of the advanced material listed under **More Activities** and **Learning Resources.** How much more to explore is only limited by your time, interest, ingenuity and motivation!

Geocases on the Web

Supplementing the printed cases in *GeoSystems Today: an Interactive Casebook* is an extensive Web site located at the following URL:

http://www.wiley.com/college/wave/geocases

The web-based materials are designed to serve learners and instructors who are users of Wiley textbooks in the geosciences: geography, geology, environmental science, earth system science and anthropology or area studies. The Web site will keep users current with new developments in the field, will give access to the most up-to-date data for lab activities and exercises, and provide a mechanism for users to interact with the case contributors—users can also propose new activities or contribute their own materials. Our hope is that this project will help set a new standard for support of textbook users. Most importantly, we hope this project will demonstrate the potential of "hybrid" pedagogical resources—publications that include traditional print media as well as interactive graphics and other visualizations that are normally too expensive to print. Again, Wiley recognizes that learners and teachers have a diversity of resource needs and learning styles—please let us know how well the print and Web site materials work for you.

The Web materials are organized under the following three major headings: first there is a brief **Case Abstract** (of the print material) which is followed by **Virtual Tours and Activities** and **Learning Resources**.

Virtual Tours and Activities on the Web

The tours are interactive slide shows that will take you virtually to the places discussed in the cases. As such they could also serve as useful regional/geographical summaries in physical or cultural geography. The purpose is to explore interactively and more in-depth via full-color maps, photos, satellite images and other visualizations the issues and places in question. The virtual tours also provide hotlinks to related materials on the Internet that complement the cases. At the end of each **Virtual Tour** are **Self-Study Exercises** that may be assigned as out-of-class activities or used in a laboratory setting.

Learning Resources on the Web

The web-site also includes an extensive set of supplementary learning resources for use by students or instructors as a starting point for specialized projects, classroom presentations, home-work assignments or supplementary reading. The learning resources are organized under the following categories:

Contact an Expert
Under this category you will find names and contact information of researchers. policy-analysts and other experts or organizations involved with the topics discussed in the cases.

Find the Facts
This category provides links to online sources of data and statistical information of various types as well as lists of important organizations.

Glossary and Key Terms
Here you will find links to existing online **glossaries** as well as an extensive list of **geoscience key terms** used in both the print and online versions.

Graphics Gallery
This section provides links to web sites with online sources of photos and

graphics related to the region or issue in question. This is in addition to those graphics included in the virtual tours.

Instructors Integrator
Here you will find pedagogical notes to instructors, including: information on how to promote active learning in the classroom, recommended prior readings, pointers on adaptation of the materials to specific courses or curricula, correlation with the national science standards or benchmarks, answer keys to the self-test exercises, and instructions for feedback to the case study contributors and for possible material contribution by you the user.

Join a Discussion Group
This section provides links to electronic discussion groups on related topics from the region or the case problem in question.

Read More About It
Supplementary reference sources, readings, and bibliographies are located in this section.

Videos and Films
A short list of videos and films that you can order is found here as well as information on where and how to acquire them.

WWW Links to Related Resources
This resource organizes a select list of URLs on the Internet covering a range of topics that complement the case. Topics include such issues as global change, ecology, anthropology, sustainable development, politics, environmental science, geography and policy-analysis. This list will be frequently culled, evaluated and updated. Included are links to projects that have online data and imagery useful for more advanced image analysis, computer modeling, GIS or remote sensing.

About the Contributors

Each case study contributor is a **specialist on the issue and place in question** and all have done field studies on the topic/region. Most of the graphical material and photos come from the contributors themselves (or their colleagues) and reflect fieldwork and lab analysis that has been published in scientific journals and books. Each contributor continues to work on the topic and will periodically **add new findings** to the online case material or indicate where you can acquire newly released publications. In some instances the case reflects a synthesis from many different scientists and writers.

You are encouraged to learn more about each contributor by going to their online biographical sketch which also has a link to their personal homepage. All the case contributors would welcome feedback on how to improve the material. A brief description of each author's affiliation and research interests is presented below:

Mark Finco, Mark Finco is currently a Senior Remote Sensing Analyst working with the USDA Forest Service's Remote Sensing Application Center (RSAC). He recently received a doctrate from the University of Utah in Geography, where he specialized in using remote sensing and GIStechnologies as they relate to environmental modeling.

Robert E. Ford is Associate Professor and Adamson Chair in International Studies at Westminster College of Salt Lake City. He is an earth systems scientist who has focused on "nature-society" interactions in many parts of the world. The Virunga case material is based upon fieldwork carried out over a four-year stint in Rwanda and the East African rift region. He has also worked extensively in semi-arid regions such as the Sahel in Africa as well as in North and Central America.

George Hepner is a Professor of Geography at the University of Utah and active researcher in the Southwestern Center for Environmental Studies. He is the Principal Investigator for several research projects that use GIS and other earth system science tools for modeling environmental hazards and vulnerability and risk along the US-Mexico border and elsewhere in the world. In addition, Dr. Hepner is an editor for the global change journal "Earth Interactions."

Michael W. Hernandez is a doctoral candidate in geography at the University of Utah in Salt Lake City. His current research focuses on risk assessment of natural hazards in rapidly urbanizing environments.

James D. Hipple is an Assistant Professor of Geography at the University of Missouri-Columbia where he specializes in remote sensing, GIS, and spatial analysis. He has been particularly active in using remote sensing to model land use and land cover dynamics in urban settings and its impact on ecological systems.

Robert Moye is a freelance geological and environmental consultant who works extensively in areas of environmental remediation and planning as well as mineral exploration and production. His work has taken him to many areas of the world including the Overthrust Belt and Persian Gulf.

W. Scott White, a doctoral candidate from the University of Utah, is an Instructor at Weber State University in Ogden, Utah. The focus of his research is on the cartographic visualization of drainage basins and other hydrological-climatic processes. Recently he has been involved with a large federally funded project studying the consequences of the 1993 floods in the Upper Missouri-Mississippi drainage basin.

Sandra A. Zicus is an environmental educator living in Salt Lake City. She works with many community organizations both in the US and abroad dealing with issues such as wetlands conservation, migratory bird protection, and outdoor science education.

Case 1 — Earthquake Hazards - The Wasatch Fault

Questions Explored:

What is the Wasatch Fault and where is it located from a global, regional, and local perspective?

What are the physiographic, geologic, and tectonic characteristics of the Wasatch Fault?

What other hazards do surface fault rupture-producing earthquakes pose along Utah's Wasatch Range?

When will the next major earthquake along the Wasatch Fault occur?

How do we coexist with earthquake hazards and mitigate the danger from a large earthquake?

Key Learning Outcomes:

The learner will be able to:

Understand the tectonic process which created the Wasatch Fault.

Describe the Wasatch Fault at different spatial scales.

Define the different characteristics of the earthquake-induced natural hazards related to movement along the Wasatch Fault.

Understand the general methods in earthquake forecasting and how the Wasatch Fault estimates were derived.

Comprehend the human vulnerability to seismic hazards along the Wasatch Fault and our roles in earthquake mitigation.

OVERVIEW

Location

This case study explores the complex and dangerous interaction between burgeoning human activity and the immense forces produced by earthquakes along the **Wasatch Fault (WF) in Utah**. The WF is really a <u>zone</u> of major fault segments comprising one of the most active and longest normal fault zones in the world. Population growth in several communities adjacent to the WF (i.e., Ogden, Salt Lake Valley, and Provo) has significantly increased over the previous five years, resulting in the building of large subdivisions and related infrastructure at an alarming pace! Development is occurring in areas exposed to the seismic hazards such as surface fault rupture and landslides. People need to be aware of the hazards around them and the risks they face when living in such a geologically active area.

Author:

Michael W. Hernandez
University of Utah

Primary Objectives

The primary objective of this case is to show how large surface fault rupture earthquakes have impacted the landscape along Utah's Wasatch Front, specifically, the formation of the WF, and how future earthquakes will affect the region's populace. Tectonic forces resulting in earthquakes are part of the system known as the lithosphere. We will examine the WF and other seismic hazards from this perspective.

Why is it important for us to study the causes and effects of earthquakes? For those of us who live in tectonically active regions such as the Wasatch Front, it's critical because we are exposed to earthquake hazards every day and are at risk of being injured or killed as well as our homes and businesses being damaged or destroyed when an earthquake does occur. More people are moving into the communities surrounding the WF every day, further concentrating the population and exposing more of them to the seismic threat. Many may not be aware of the potential dangers they are exposed to by living near the segments of the WF. Increasing public awareness of earthquake dangers and earthquake preparedness are key to saving lives and reducing damage. Information on where you can get the facts about earthquakes and how to protect yourself is provided to help you prepare for such an event.

The key questions for us to examine are the following: With a high probability of an earthquake occurring along a segment of the WF within our lifetime, can we adequately forecast the next large earthquake? Also, how do we mitigate the dangers of the associated seismic hazards so we can coexist with nature's most formidable landscape altering process?

BACKGROUND

What is the Wasatch Fault and where is it located from a global, regional, and local perspective?

Global Perspective

The vast majority of the world's earthquakes occur at plate boundaries. However, large intraplate earthquakes do occur, as evidenced by the New Madrid earthquakes in Missouri (1811-12). The WF is another product of significant intraplate seismic activity. It is located within the western interior of the North American Plate.

Regional Perspective

Within the western interior of the North American Plate, the WF is located in the Intermountain Seismic Belt (ISB), a narrow region of frequent seismic activity and a volcanic "hot spot" in Yellowstone National Park (Figure 1).

The ISB is a N-S trending zone of historical seismicity that extends 800 miles from northern Arizona to Idaho and Montana and occurs within the Basin and Range province. The eastern boundary of the ISB is the transition zone from the Basin and Range province (also called the Great Basin) to the west and the Middle Rocky Mountains province (Wasatch Range, Uinta Mountains, and the northern Colorado Plateau) to the east. The Basin and Range consists of North/South-trending, regularly spaced elongated mountain ranges separated by broad basins filled with eroded sediment. A change in the North American plate movement created extensional forces which resulted in crustal rifting and characteristic block faulting, producing high-angle normal faults and creating **horsts and grabens** (Figure 2). The WF is a high angle normal fault zone created by these tectonic forces. It forms the prominent west-facing escarpment of the Wasatch Front, extending from just across the Utah/Idaho border south to Fayette, Utah.

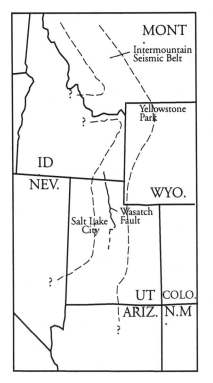

Figure 1. The Intermountain Seismic Belt showing the Wasatch Fault. Note the location of Yellowstone Park, which is situated above an active volcanic hot spot (modified from Arabasz et al., 1992).

Evidence for dynamic tectonic activity in the region is seen in both the active seismic record for the ISB and the presence of a volcanic hot spot in southwest Wyoming's Yellowstone National Park. The Yellowstone Caldera is active today, as evidenced by numerous small earthquakes, ground deformation, and hydrothermal events. Yellowstone continues to be a sleeping giant, stirring every now and then to keep us humble!

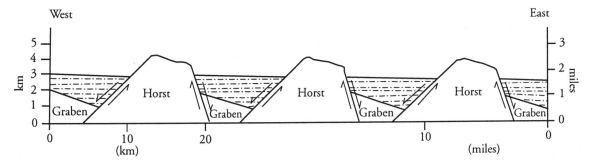

Figure 2. Typical East-West cross section of a portion of the Basin and Range province. Stippled areas are valley-fill deposits and arrows indicate relative movement of block faults (modified from Hecker, 1991).

Local Perspective
The Wasatch Fault is essentially a North-South oriented major normal fault system approximately 343 km (213 miles) long, extending from Malad City, Idaho to Fayette, Utah (Figure 3).

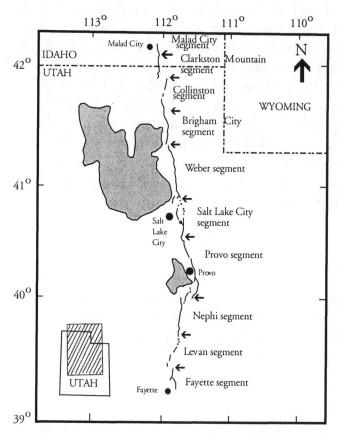

Ten major discrete segments comprise the WF. These segments behave independently during large, surface-rupturing earthquakes and all have been associated with such earthquakes since the Late Pleistocene. The five central segments (Brigham City to Nephi) have had two or more surface-fault rupture earthquakes in the last 6,000 years! This is signifcant because greater than 80 percent of Utah's population lives in close proximity to the five central segments, where earthquake risk is the greatest in the entire state! The remainder of this study focuses on this high risk zone.

Figure 3. The Wasatch Fault divided into its 10 major segments. Segment names are the boldfaced text to the right of the arrows pointing to the segments (modified from Machette et al., 1992).

What are the physiographic, geologic, and tectonic characteristics of the Wasatch Fault?

How were the Wasatch Fault segments identified? In some cases, someone saw a distinct feature on the ground or from above, looking out of an airplane window or at an aerial photograph or satellite image, which was then investigated in the field and determined to be a surface expression of a fault. Along the Wasatch Fault, several distinct surface expressions are seen in both photographs (ground and aerial) and satellite images. **Fault scarps** along many of the WF segments are visible at numerous locations. Each surface-faulting earthquake can produce a scarp up to 6.1 meters (20 feet) high!

Triangular facets can also be easily seen in both photos and satellite imagery. They are remnants of the fault plane created by the uplifting of the Wasatch mountain block. The triangular shape is formed by accelerated stream erosion in the drainage valleys caused by uplift of the mountain block.

In the field, **slickensides** (polished surfaces) and **fault gouge** (ground-up material) are key indicators of an exposed fault plane (surface). At the Salt Lake Segment just north of Salt Lake City, past quarrying for building materials exposed a fault surface. A photograph showing both fault gouge and slickensides, with their characteristic etches and grooves in the polished and striated rock marking the actual uplifted fault block surface, can be viewed in the accompanying *Virtual Tour*.

Finally, **springs** are commonly associated with the Wasatch Fault segments. The faults provide a conduit for mineral-rich groundwater to migrate to the surface, forming pools of water surrounded by dry rock. Characteristic features to look for in locating springs are a narrow band of phreatophytes (plants which only grow in water-saturated conditions) such as cattails, and warm water with a very strong odor due to its high mineral content.

What hazards do surface fault rupture-producing earthquakes pose along Utah's Wasatch Range?

Surface Fault Rupture

Along each segment of the WF, surface fault rupture is the most noticeable hazard related to fault movement from earthquakes. Surface fault rupture results in near vertical displacement of the Wasatch mountain block upward and the valley block downward along a normal fault. Paleoseismic studies in Utah have determined that the magnitude threshold for surface fault rupture in Utah is approximately 6 to 6.5. Historical seismicity records from the Basin and Range Province and the Intermountain Seismic Belt were used to model the maximum earthquake size expected in the Wasatch region - about a magnitude 7.5 earthquake. Trench studies along several Wasatch Fault segments concluded that maximum displacement along a fault segment during a single earthquake event is estimated at upwards of about 6 meters (20 feet), with an average of 1.8 to 2.4 meters (6 to 8 feet)! That's enough movement to completely destroy most structures built across the fault trace or very near it! A <u>zone of deformation</u> occurs on the downthrown valley block. It consists of small cracks, tilted blocks and possibly an **antithetic fault** scarp forming a small graben between the two fault scarps and dipping east towards the mountain block. The zone of deformation can be up to several hundred feet in width and is not a very stable place to live! The length of the fault segment is important because it places constraints on the magnitude of potential earthquakes along the WF (i.e., why earthquake magnitudes are not >7.5 along the WF).

Other Seismic-Related Hazards

Many other earthquake related hazards will have a significant impact on the areas surrounding the WF. Many moderate (i.e., Richter magnitude 5.0), non-surface fault rupture earthquakes are more common in Utah. These types of quakes can create ground motions capable of immense destruction! Moderate 5.0+ earthquakes (which have a 10-year recurrence interval along Wasatch Front region) can potentially cause damage from **ground shaking**, which initiates landslides and rockfalls, causes flooding, and eliminates soil stability through soil liquefaction. Many of these hazards can occur up to 80 km (50 miles) from the epicenter depending on the earthquake magnitude and geology!

Ground shaking, like surface fault rupture, is a hazard DIRECTLY related to the energy released by an earthquake. Ground shaking is the principal cause of loss from earthquakes! Ground shaking is caused by the movement of seismic waves through the upper layers of surface rock (bedrock and regolith), causing vibration of the ground, producing small short term lateral and vertical ground displacements. Seismic waves

11

travel more efficiently through crustal hard rock than soft, unconsolidated sediments and therefore, can be felt at much greater distances from the epicenter. However, soft, unconsolidated sediments tend to AMPLIFY seismic waves resulting in enhanced ground shaking (greater ground displacement) and much greater loss. Adjacent to the Wasatch Front, most of the valley floors immediately west of the central five Wasatch Fault segments (between Ogden and Provo) are filled with unconsolidated lake and alluvial fan deposits (gravel, sand, clay). This is BAD NEWS! Enhanced ground shaking is a real threat to these areas. Only buildings constructed directly on bedrock are protected from the amplified ground shaking effects!

Soil liquefaction is a secondary earthquake hazard caused by ground shaking, which turns stable, solid water-saturated sediment into a liquefied unstable mass resulting in loss of foundation support. A shallow water table is usually also present. Liquefaction potential depends on (1) soil and groundwater conditions, (2) duration of ground shaking, and (3) an earthquake with strong enough ground shaking (magnitude > 5.0). Along the WF, the highest liquefaction potential occurs in areas with shallow groundwater (<30 feet) and loose, sandy soils, where these soils can become easily saturated with water. These conditions are widespread along the central part of the WF, west of the fault segments.

Slope failures are also a secondary earthquake hazard caused by ground shaking. Because the Wasatch Fault is along a mountainous front, loss from slope failures is expected to be high in these areas. Historical records indicate that earthquakes with magnitudes of 7.5 or greater can cause slope failures up to 300 km (185 miles) from the epicenter! Between 1850 and 1986, 12 earthquakes in or near Utah (magnitude range 4.3 to 6.6) resulted in slope failures. There are several types of slope failures that are caused by earthquakes along the Wasatch Front: landslides, debris flows, and rock falls.

Earthquake induced <u>landslides</u> occur in regions with steep slopes, when ground shaking from magnitude 4.5 or greater earthquakes can cause earth material to break loose and slip towards the base of the slope. <u>Debris flows</u> are a variable mix of soil, water, rock, and vegetation forming a moderately thick slurry that flows downslope as a pulse or a surge due to gravity. Areas susceptible to debris flows occur (1) near the top of the watershed, (2) have steep slopes and highly weathered bedrock, and (3) enough water (i.e., from cloud bursts or mixture of material with stream water) to create a flow. <u>Rock falls</u> occur when ground shaking from earthquakes with magnitudes as low as 4.0 causes newly detached rock clasts to free-fall and/or roll from a very steep slope or cliff at a relatively high velocity. They are the most common earthquake-induced type of slope failure. A large-scale earthquake (magnitude 7 to 7.5) along the WF is expected to produce thousands of rock falls within the Wasatch Mountains! The greatest potential for loss is in the bench areas along the Wasatch Front and the mountain canyons., where housing developments have been built in the rockfall runout zones.

Earthquake-related **flooding** can be a secondary or tertiary hazard in the Wasatch Front region. Flooding can be caused by rupturing of a mountain reservoir or dam from ground shaking or the result of <u>tectonic subsidence</u> – the deformation, tilting, and lowering of valley floors associated with a large surface fault rupturing earthquake generated along a normal fault like the WF. Land adjacent to lakes (i.e., the Great Salt Lake) and reservoirs or land with a shallow groundwater table would be inundated. Along the WF, the estimated area of subsidence extends about 10 miles west of the Wasatch Fault segments, with the majority of deformation and maximum subsidence occurring within three miles of the fault. Floods can also be caused by earthquake-generated **seiches**, large waves produced in lakes or reservoirs that suddenly arise to great heights and break along the shoreline. The waves produced by the seismic-generated oscillations act like sloshing water back and forth in a shaken bucket. Earthquake-generated seiches in the Great Salt Lake can be 12 feet or more in height. The highest risk from seiches along the central five segments of the WF are the shorelines along the Great Salt Lake and Utah Lake near Provo.

Finally, flooding can also be a tertiary hazard generated by other secondary earthquake-generated hazards such as landslides (dam a stream upslope which eventually fails) and ground tilting (causes subsidence at one end of tilted block where shallow groundwater or stream flow collects). The key factor is the availability of a water source to cause flooding, such as the rupturing of water tanks, reservoirs, pipelines, and aqueducts or the blockage or diversion of a stream course.

THE HUMAN DIMENSION

When will the next major earthquake along the Wasatch Fault occur?

To reduce the severity and extent of loss to the infrastructure and human lives, we must be able to (1) better forecast the probability of large earthquakes and (2) develop effective earthquake hazard mitigation strategies for our communities. Earthquake forecasting attempts to predict both where and when earthquakes will occur and at what intensity and duration they will affect the surrounding region. The ultimate goal is to be able to pinpoint the date and time an earthquake will occur and thus provide effective and accurate early warning to the population to minimize human loss.

Long-term forecasting, which is based on understanding the tectonic cycle, has been somewhat successful in predicting large earthquakes within a broad time period. It applies a probabilistic approach to predicting earthquake activity. Accurate forecasting requires interpreting paleoseismic (pre-historical) data along each fault segment. This is done by trenching across the fault scarp and interpreting the relationship between sedimentary features to determine the number of earthquake events that have occurred along that fault segment. The key to determining the earthquake recurrence interval is to accurately date each group of sedimentary features in the trench. In Utah, both the radiocarbon and the thermoluminescence techniques have proven to provide reliable dates. Nineteen surface faulting earthquakes occurring over the last 6,000 years were identified through trench studies in the central WF segments (Brigham City to Levan). Across the WF, the composite recurrence interval is approximately every 350 years, with the last surface-faulting earthquake occurring approximately 600 years ago along the Provo segment. Therefore, statistics indicate that the central WF segments are in a **seismic gap**, which means a large surface-faulting earthquake could occur at anytime, making implementation of earthquake mitigation strategies paramount!

Conversely, **short-term earthquake forecasting** tries to pinpoint the exact time and location of earthquakes, but successful predictions have not been the norm. Scientists apply a deterministic approach, relying on observations of <u>earthquake precursors</u> - anomalous physical changes on the earth that are warning signs of seismic activity. Changes in (1) ground bulging or tilting, (2) rock magnetic field, (3) electrical conductivity of the rock, (4) water well levels, (5) amount of radon in well water, (6) unusual animal behavior, and (7) foreshocks are some of the precursors used in predicting earthquakes. An earthquake was successfully predicted for Haicheng, China in 1975 based on numerous foreshocks, slow tilting of the land, and magnetic field fluctuations. More than one million people were evacuated before the magnitude 7.3 earthquake hit, resulting in the deaths of only a few hundred people. However, some large surface faulting earthquakes did not provide any evidence of precursor activity prior to their occurrence. Short-term prediction is therefore, difficult at best because we cannot see the actual mechanisms and processes causing earthquakes working underground, and without precursors, we do not have any specific warning until it is too late! Newer techniques are being developed to measure very small changes in the surface of the ground using GPS (global positioning systems) and multi-date satellite radar change detection. The data help scientists refine their understanding of the deformation pattern caused by earthquakes. This can lead to better earthquake prediction based on observations of the changes in the rate and extent of crustal deformation. Also, the installation of seven more ground-motion detectors in the ISB has provided scientists much better data to decipher precursor activity of Utah earthquakes.

We know large surface faulting earthquakes will occur again, so what can we do to protect ourselves? Choosing to live along the Wasatch Front requires us to accept some earthquake risk - it is all around us! We cannot control nature but we can strive to understand it and prepare for the consequences as best we can by applying proven mitigation strategies to reduce the risk of loss from earthquake activity.

How do we coexist with earthquake hazards and mitigate the danger from a large earthquake?

Earthquakes will always be a danger to communities that exist within seismically-active regions. We have moved into potentially unstable areas which make us vulnerable to earthquake hazards. It is impractical to think we could move an entire city away from the WF or that we could totally eliminate the earthquake hazards. So how do we co-exist with earthquakes? The first step in reducing earthquake risk is to make the public aware of both the potential risks from earthquakes as well as the critical steps required to reduce loss. People cannot make an informed decision about how much risk they are willing to accept unless they are made aware of what earthquake hazards might affect them! Currently, scientists can artificially induce small earthquakes through deep-well fluid injection, reservoir filling, and underground nuclear explosions, which can poten-tially reduce built-up strain in rock. However, there is no proof that this action will prevent large earthquakes from occurring. Therefore, other mitigation steps must be taken to "reduce" our vulnerability to earthquake-related hazards.

Scientists and planners need to address six critical issues to mitigate earthquake-related hazards in vulnerable communities. First, accurate hazard maps need to be completed and used for effective land use planning in areas not already developed. This is the most effective mitigation strategy because it ensures the land is properly used in relation to the hazards that affect it! Along the WF, this might mean not allowing further heavy development in high liquefaction potential areas, where enhanced ground shaking will be most prevalent. In the 1989 Loma Prieta, CA earthquake, 70% of the total property damaged was due to enhanced ground shaking. Also, statewide fault setback ordinances were implemented to ensure no development occurs within specified distances on either side of the fault scarp, while allowing appropriate low risk land use activities such as farms, golf courses, open space, and parks to exist along the faults.

The second critical issue requires the evaluation of existing development for earthquake resistance and reinforcing those structures that are expected to collapse or be severely damaged during a large earthquake. Most people killed or injured by an earthquake were in buildings or homes which collapsed or broke apart, dropping pieces of ceiling or walls on the occupants or catching fire. Creation of an earthquake-resistant building code is essential to ensure new building sites meet the required geologic stability criteria and are constructed using approved structural designs and flexible material that will not crumble and fall on the occupants. Utah's Uniform Building Code was implemented in 1987 but needs refinement as better earthquake models for the WF have been developed based on enhanced seismic data for the region.

Communities in earthquake-prone regions need to develop and continually refine comprehensive emergency planning and response programs to ensure all government agencies and utilities know how to prepare for and respond to a large earthquake. This will ensure that when normal utilities, transportation routes, and communications are interrupted, agencies will know how to respond to emergencies effectively and get aid to those in need as well as getting services back on-line in minimal time. In Utah, the Division of Comprehensive Emergency Management has developed these plans which are initiated through their operations center behind the Utah State Capitol building. A recovery plan should also be developed to ensure the services and infrastructure are rebuilt in order of established priority to bring the community back to normal.

Finally, the most important critical issue to the average citizen - YOU - is public education on what you and your family should do before, during, and after an earthquake. There is a wide variety of information on earthquake preparedness that has been published and is available to everyone. Some good internet web sites and publications relating to earthquake preparedness are listed in the RESOURCES section of Case 1, found on the home page of *GeoSystems Today: An Interactive Casebook*.

What tectonic processes might be causing the seismic activity found in the Intermountain Seismic Belt and has resulted in the formation of the Wasatch Fault segments? Where could you find information on seismic activity and earthquakes where you live?

How might you recognize the surface expression of a fault in your community? Where could you get information on faults and other earthquake-related hazards?

Generally, are you safe from all the effects of a large-magnitude surface fault rupture earthquake if you live within the same valley of the earthquake, 20 km (12.5 miles) from the epicenter? If not, what hazards might you expect to encounter from the earthquake?

After learning about many of the earthquake hazards associated with the Wasatch Fault, would you be willing to live in a seismically active region such as the Wasatch Front? Why or why not?

SOME RECOMMENDED READINGS

Books:

Drury, S. A.. 1993. *Image Interpretation in Geology* (2nd Edition). London: Chapman & Hall.

Murck, Barbara W., Brian J. Skinner, and Stephen C. Porter. 1996. *Environmental Geology.* New York: John Wiley & Sons, Inc.

Murck, Barbara W., Brian J. Skinner, and Stephen C. Porter. 1997. *THE DANGEROUS EARTH: An Introduction to Geologic Hazards.* New York: John Wiley & Sons, Inc.

Nuhfer, Edward B., Richard J. Proctor, and Paul H. Moser. 1993. *The Citizens' Guide to Geologic Hazard.* Arvada, CO: The American Institute of Professional Geologists.

Skinner, Brian J. and Stephen C. Porter. 1995. *The Blue Planet: an Introduction to Earth System Science.* New York: John Wiley & Sons, Inc.

Smith, Keith. 1996. *Environmental Hazards: Assessing Risk and Reducing Disaster* (2nd Edition). London: Routledge.

Stokes, William Lee. 1986. *Geology of Utah.* Utah Museum of Natural History, University of Utah. Occasional Paper No. 6.

Strahler, Alan and Arthur Strahler. 1996. *Introducing Physical Geography: Environmental Update.* New York: John Wiley & Sons, Inc.

Other Publications:

Arabasz, W.J., J.C. Pechmann, and E.D. Brown. 1992. Observational seismology and the evaluation of earthquake hazards and risk in the Wasatch Front area, Utah. In Gori, P.L., and W.W. Hays, eds. *Earth-*

quake Studies in Utah. U.S. Geological Survey Professional Paper 1500-D.

Black, Bill D., William R. Lund, David P. Schwartz, Harold E. Gill, and Bea H. Mayes. 1996. *Paleoseismic investigation on the Salt Lake City segment of the Wasatch Fault Zone at the South Fork Dry Creek and Dry Gulch sites, Salt Lake County, Utah.* Paleoseismology of Utah, Vol 7. Utah Geological Survey Special Study 92.

Eldredge, Sandra N. 1996. *The Wasatch Fault.* Utah Geological Survey Public Information Series 40.

Eldridge, Sandra N. 1992. *Places With Hazards: A teacher's handbook on natural hazards in Utah for secondary earth science classes - geologic hazards lecture set: the earthquake hazard in Utah.* Utah Geological Survey Open-file Report 211 – A.

Christenson, Gary E. 1994. *Earthquake Ground Shaking in Utah.* Utah Geological Survey Public Information Series 29.

Hecker, Suzanne. 1993. *Quaternary Tectonics of Utah with Emphasis on Earthquake-Hazard Characterization.* Utah Geological Survey Bulletin 127.

Jackson, Michael. 1991. *The number and timing of Holocene paleoseismic events on the Nephi and Levan segments, Wasatch Fault Zone, Utah.* Paleoseismology of Utah, Vol 3. Utah Geological Survey Special Study 78.

Lowe, Mike. 1990. *Geologic hazards and land-use planning: background, explanation, and guidelines for development in Davis County in designated geologic hazards special study areas.* Utah Geological and Mineral Survey Open-File Report 198.

Machette, M.N., S.F. Personius and A. R. Nelson. 1992. Paleoseismology of the Wasatch fault zone—A summary of recent investigations, conclusions, and interpretations, in Gori, P.A., and W. W. Hays, eds. *Assessing regional earthquake hazards and risk along the Wasatch Front, Utah*: U.S. Geological Survey Professional Paper 1500, Chapter A, p. A1-A72.

Shedlock, Kaye M. and Louis C. Pakiser. 1996. *Earthquakes.* U.S. Geological Survey.

Utah Earthquake Preparedness Information Center (EPICENTER). *EARTH-QUAKES: What You Should Know When Living In Utah.* Utah Division of Comprehensive Emergency Management.

Articles:

Ashland, Francis X. 1997. New strong-motion instruments will monitor earthquake ground shaking in Utah. *Survey Notes* Vol 29 (2):6-9. Utah Geological Survey.

Christenson, Gary E. 1991. Earthquake hazards of Utah. *Survey Notes* Vol 24 (3):3-10. Utah Geological and Mineral Survey.

Case 2

The 1993 Floods on the Mississippi and Missouri Rivers

Questions Explored:

How do scientists define and characterize floods?

What were the U.S. Midwest climatic conditions during the spring and summer months of 1993?

How can geographic technologies such as remote sensing, GPS, GIS, and GVIS aid in modeling flood hydrology?

What role did the human-altered landscape play in exacerbating the flood?

What lessons were learned by federal, state, and local government agencies after the floodwaters receded?

Key Learning Outcomes:

The learner will be able to:

Identify on a map the location of the Upper Mississippi River Basin and the part of the United States most affected by the 1993 Midwest Flood.

Understand and discuss the climatic conditions leading up to the flood and the precipitation patterns during the flood.

Understand how the results of Geographic Information Analyses can be used to visualize the flood and its effect on the landscape.

Understand the flood's effect on human settlement patterns, agriculture, and government policies.

OVERVIEW

Location

In this learning activity, you will investigate the large flood that occurred over the spring and summer months of 1993 in the **Upper Mississippi River Basin** (UMRB). Researchers have defined that portion of the UMRB that was affected by the 1993 floods as a 700,000 km² (~270,000 mi²) region spread over nine Midwestern and Great Plains states (Figure 1). This part of the basin encompasses both the main stem of the Mississippi River above Cairo, Illinois, as well as the main stem of the Missouri River below Gavins Point Dam near Yankton, South Dakota. It also includes the numerous tributaries of these rivers such as the James, Des Moines, Illinois and many other rivers and streams. A total of 532 counties in the Midwest (including all of the counties in Iowa) were considered Federal disaster areas as a result of the flood.

Author:

W. Scott White
Weber State University

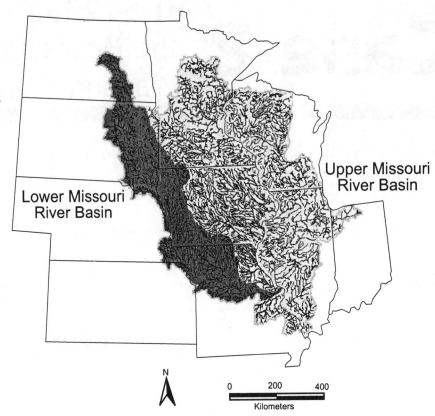

Figure 1. The Upper Mississippi River Basin (UMRB) affected by the 1993 floods is spread over nine Midwestern and Great Plains states.

Primary Objectives

There are three primary objectives in this case study. The first objective is to acquaint the readers with flooding as part of the hydrologic cycle. This section will describe the natural process of flooding and the water balance, ending with a description of the flood hydrograph. The second objective focuses specifically on the **Midwest Flood of 1993**, and includes a brief chronology of pre-, peak-, and post-flood events, and a description of some other **historic Mississippi River floods**. The section will end with an examination of the **computer methods** scientists use to predict and monitor large-scale floods. The third objective deals with the human dimensions of the flood, including a look at the **structural and nonstructural means of controlling floods**, and their results.

When investigating the 1993 Midwest flood described in this case study, the reader should focus on several key questions: What lessons have been learned from the 1993 Midwest floods? How do certain U.S. government agencies provide disaster support and land use management information to people who live on the floodplain? What were the real costs to the nation due to the floods? Can new computer technologies aid flood forecasters and planners in preparation for future big floods?

BACKGROUND

How do scientists define and characterize floods?

A flood may occur when a stream overflows its banks, allowing water to spread over the stream's floodplain. Floods can be considered a part of the hydrologic cycle – a series of related processes that defines the motion of water through the natural system. The processes of precipitation, water flow (surface and subsurface), and erosion are all a part of this cycle. Flood researchers are particularly concerned with the water balance (sometimes called the **water budget**) which is the amount of **inflow**, **outflow**, and **water storage** that occur in an area. By examining the water balance, scientists can determine how much precipitation contributed to surface water and groundwater flow (inflow), and how much water left an area through ground or surface runoff (outflow), or through **evapotranspiration** (the loss of water through evaporation and transpiration by vegetation). They can also ascertain how much water was stored in an area through lakes, reservoirs, groundwater, soil moisture, or snow cover.

Floods are often characterized by their **temporal and areal distribution**. The **recurrence interval** of floods is an average time period between major flood events of a certain size and magnitude. Engineers use this measurement in the design and construction of **dams** and **levees**. The calculation of recurrence intervals depends upon an accurate historical record of streamflow as recorded through a nationwide network of **stream gauging stations** operated by the **United States Geological Survey (USGS)**, state, and local agencies. The recurrence interval is actually more of a probability estimate, so that a 100-year flood has a 1/100 or 1% chance of occurring in any given year. Engineers also construct **stream hydrographs**, which are graphs of **stream discharge** per unit time. The discharge measurements come from gauging station reports, and are useful in examining the duration of a hydrologic event. **Figure 2** shows a typical flood hydrograph for a gauging station on the Mississippi River in St. Louis, Missouri. Discharge through this station peaked on August 1, 1993, at over 1,000,000 cubic feet per second (cfs) and is indicated by the bold vertical line in the figure.

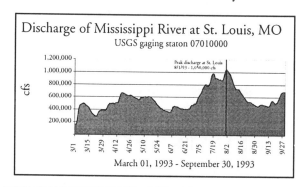

Figure 2. Stream hydrograph – USGS gauging station on the Mississippi River, St. Louis, Missouri.

What were the U.S. Midwest climate conditions like during the spring and summer months of 1993?

The flood in the UMRB during the spring and summer months of 1993 was one of the nation's most costly natural disasters. Hydrologic records that had stood for decades were broken along the major rivers and tributaries. Millions of acres of cropland were damaged, resulting in huge financial losses throughout every sector of the agriculture industry. Thousands of people were either temporarily or permanently displaced, while 52 people lost their lives due to flood-related circumstances. However, a flood of some magnitude was not totally unexpected. Climatic conditions had been pointing towards a possible disaster for months prior to the main summer flooding period.

In the fall of 1992, **soil moisture** values in the central U.S. were much higher than normal. A heavy snow season further exacerbated the situation, so that by March 1993, the soil moisture capacity had been reached for much of the Midwest and Central Plains states. Precipitation intensified in the form of heavy rains during April and May due to a **Pacific Ocean low-pressure trough** and a **high-pressure ridge in western North America** which brought westerly flow into the central U.S. From June through July, the weather patterns shifted, bringing a low-pressure trough to west central North America. Moisture from the Caribbean was brought northwards, triggering very heavy precipitation events and extensive flooding throughout the Midwest. Heavy rainfall continued intermittently through August, but by September, the low-pressure trough which had been relatively stationary over the Midwest moved east of the UMRB, causing the storms to weaken.

Were the 1993 floods in the Midwest unusual phenomena, or are there historical examples of such events? The 1993 floods equaled or exceeded flood recurrence intervals of 100-500 years along major stretches of the upper Mississippi and lower Missouri Rivers. A look back at Midwest hydrologic records from the mid-1800s to today shows that there have been a number of "**great floods**" along the upper Mississippi River. Heavy rains occurred throughout the fall and winter months of 1926 and 1927, breaking gauge records up and down the length of the river. Unlike the 1993 flood which was essentially mitigated by the time the floodwaters reached Cairo, Illinois, the 1927 flood broke through huge levees along the lower Mississippi in the spring of that year, flooding millions of acres of land. Other floods have occurred in the UMRB since 1927, including disastrous ones in 1937 and 1973, however, both floods were overall less hazardous than the 1927 event due to flood-control works that were built during the intervening years.

There has been some scientific discussion since 1993 concerning the relationship between the UMRB floods and the global weather phenomenon known as **El Niño** – a disturbance of the ocean-atmosphere system in the tropical Pacific Ocean which is believed to cause weather abnormalities worldwide. Consequences of El Niño are increased rainfall across the southern section of the U.S. and in Peru, causing destructive flooding, while drought and sometimes brush fires occur in areas of the western Pacific. Some scientists now believe that the floods may have been indirectly related to El Niño events, which influenced the development of the jet stream and the resulting spring and summer Midwest storm path. However, the intensity and persistence of the rainfall over the UMRB seems to be more directly related to the unusually strong ridge-trough pattern in the area, and the relatively common summer influx of Caribbean moisture causing summer storms in the Midwest. As we begin to understand more about El Niño's effect on regional weather systems (particularly after the 1997-98 El Niño season), the phenomenon's relationship to the 1993 floods should become clearer.

How can geographic computer technologies such as remote sensing, GPS, GIS, and GVIS aid in modeling flood hydrology?

Computers have revolutionized the way environmental scientists model earth processes. Since the advent of digital satellite imagery in the early 1970s, **remote sensing** (the study of the earth's surface using remote detection or sensing instruments not in direct contact with the surface) has played a vital role in mapping and analyzing flooded terrain. One of the primary applications of **satellite imagery** is in the mapping of **flood extent**, and the determination of flood damage. Pre-, peak-, and post-flood suites **of Landsat Thematic Mapper (TM) images** have been used to identify changes in the landscape due to the floods (**Figure 3**). The Landsat TM images have a **spatial resolution** of 30 meters, and are able to detect seven **bands of visible and near-infrared light**.

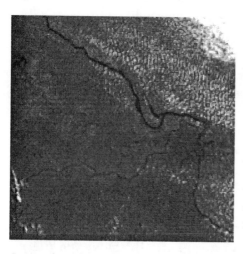 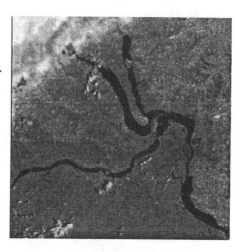

Figure 3. Landsat TM images of the St. Louis, MO area. The left image is from 7/15/92 illustrating pre-flood conditions. The right image is from 8/19/93 illustrating peak-flood conditions.

Several technologies are being used together to provide a better understanding of the effects of the 1993 floods. The **Federal Emergency Management Agency (FEMA)** conducted a study during the summer of 1993 using **Global Positioning System (GPS)** and **Geographic Information System (GIS)** computer technologies. GPS units can acquire fairly precise **geodetic coordinates** of locations on the earth's surface using a network of 24 **earth-orbiting satellites**. The real-time coordinates were then fed into a mapping and geospatial analysis database known as a GIS to get accurate and timely maps of flooded areas. GPS/GIS mapping is preferable to traditional hand drawn mapping techniques because of the time and money saved using the computers. Also, the accuracy of the GPS satellite network means that FEMA damage assessments are precisely mapped and documented.

GIS is also being used in conjunction with new, powerful scientific visualization software to create dynamic representations of the 1993 floods. The results of a GIS-based water balance study of the entire 1993 UMRB flood area are being incorporated into **Geographic Visualization (GVIS)** models. In this GVIS study, the static results of the GIS (a series of daily water storage maps for the flood period) will be transformed into a series of animated maps showing the pulses of water movement through the basin. GVIS promises to bring a greater visual understanding of the flood to both researchers and the general public, thus increasing our capacity to deal with future floods in the UMRB and elsewhere. A large archive of flood-related GIS data sets is available through the **Internet** home page of the **Scientific Assessment and Strategy Team (SAST)**, an intergovernmental organization formed in November, 1993, to study the floods.

What role did the human-altered landscape play in exacerbating the flood?

One of the most important questions that scientists and engineers asked themselves after the flood was to what extent did human modification of the UMRB river system worsen the effects of the flood. A primary engineering method of flood control is **channelization**, which seeks to increase the channel's cross-sectional area by increasing stream depth and width, or by straightening the channel. These processes should allow a stream to handle a greater discharge of water at higher velocities, however, channelization has a negative side. Many people believe that channelization destroys natural habitats and ecosystems, causes surface and groundwater pollution, and ruins the aesthetic value of a stream. In some cases, intense channelization may control flooding locally, but contribute to more intense flooding further downstream. Another problem with channelization is that any records of stream discharge gathered along a modified stream channel are not valid in a historical context. Predicted flood stages based on historical streamflow records cannot be trusted completely because of the change in channel morphology. **Flood stage predictions** made during the 1973 and 1993 floods were only rough estimates based on gauge data from channelized streams.

The great farmlands of the Mississippi and Missouri River basins would not be possible without the help of manmade levees and drainage systems designed to protect croplands from relatively small floods (~25 year or less return period). However, the 1993 floods damaged many levees along these rivers through **overtopping** and **breaching**. Did the levees in some way contribute to the floods' intensity? A 1995 study by the U.S. Army Corps of Engineers (USACE) utilized hydraulic models developed for individual reaches of the Mississippi and Missouri Rivers, and several tributaries. The study showed that flood stage was reduced when all agricultural levees were removed from a floodplain. Under this scenario, the stage of the Mississippi River above St. Louis would have been lowered by an average of 1.0 m, with an 0.6 m decrease in stage on the lower Missouri River. The greatest increases in flood stage occurred when the scenario called for raising all agricultural levees to a level that 1993 floods would not overtop them. Stages were about 0.9 m higher on the Mississippi River and 1.1 m higher on the lower Missouri River. High rates of stream discharge did not always correlate with higher flood stages in the USACE study, leading one to conclude that no single floodplain management scenario can control all of the flooding along a major river reach.

What lessons were learned by federal, state, and local government agencies, after the floodwaters receded?

Prevention, mitigation, and support are key components of government policies dealing with natural disasters such as the 1993 Midwest floods, but did the policies work during and after the floods? How did policies change as a result of the floods? To answer these questions, we need to turn to a brief study of the lessons learned from the floods.

Some of the most obvious lessons learned during the floods were that **engineered structures** did not always hold the water in the river channels as they were designed to do. Levees along many UMRB watercourses failed and were either breached or overtopped by floodwaters. These structures were generally designed to hold back floodwaters from much smaller flood events, thus their failure is not that surprising. Some people have blamed the levees for worsening both upstream and downstream Mississippi River flooding by raising the heights of the river to extreme flood levels.

Since 1993, many flood experts have begun to look at **nonstructural means of reducing the effects of floods.** Most of these approaches are geared towards **floodplain zoning and land use planning.** Coupled with engineered systems such as levees and dams to protect preexisting urban areas, wise use of floodplains through development limitations is becoming more commonplace. As a result of early 1990s disasters such as floods and hurricanes, FEMA has undertaken a more substantial role in preparing the nation for natural and human-caused disasters and responding to these disasters through education, zoning and planning regulations, and monetary assistance. FEMA is considered as a supplemental source of relief after a disaster overburdens state and local governmental resources.

Of all the lessons learned from the floods, the most important one was the fact that the people living in the UMRB area were not prepared for a **500-year event** such as the one which occurred in 1993. This chain of responsibility goes from the federal government to the farmer living and working on the floodplain. Problems identified after the floods included those relating to flood monitoring equipment and procedures, public information handling and dissemination, and land use planning. Mitigation projects that had worked in the past failed or were only marginally successful during this flood. Policies at the local, state, and federal governmental levels are being reviewed and changed, so there were some benefits as a result of the floods. It is hoped that continued support of these government agencies for flood research will lead to better flood prevention and mitigation policies before the next major flood in the Upper Mississippi River Basin occurs.

The 1993 floods were undoubtedly a costly disaster to the U.S. However, "cost" can describe many things: money, time, energy, damage, death, etc. The real cost to the American public of the Midwest floods was perhaps a sense that we cannot fully control nature, no matter how many dollars are spent to prevent such a disaster. Most lessons are learned the hard way, but it is through such events as the 1993 floods that we become a stronger nation, able to face the challenge. Through scientific research, governmental assistance, and public cooperation, the challenges of preventing another costly flood disaster are being addressed.

What was the size of the flooded Upper Mississippi River Basin? How much of the land area of the conterminous U.S. was affected by the floods?

What parts of the hydrologic cycle are affected by man's alteration of the natural landscape? Could this affect a region's water balance measurements?

How did the 1993 floods compare in magnitude to earlier Mississippi River basin floods? Did media hype make the floods appear worse than they really were?

How have previous El Niño years compared to the relatively minor El Niño-like weather patterns during 1993?

Should channelization and levee construction be integral parts of U.S. flood prevention policy? What other alternatives (if any) are available?

SOME RECOMMENDED READINGS

<u>Books</u>:

American Meteorological Society, 1994. *The Great Flood of 1993*. American Meteorological Society.

Beven, Keith and Paul Carling (eds.). 1989. *Floods: Hydrological, Sedimentological and Geomorphological Implications*. John Wiley and Sons.

Changnon, Stanley A. (ed.). 1996. *The Great Flood of 1993*. Westview Press.

Clark, Champ. 1982. *Flood*. Time-Life Books.

Galat, David L. and Ann G. Frazier (eds.). 1996. *Overview of River-Floodplain Ecology in the Upper Mississippi River Basin*. v. 3 of Kelmelis, J.A. (ed.). *Science for Floodplain Management into the 21st Century*. U.S. Government Printing Office.

Griggs, Gary B., and John A. Gilchrist. 1983. *Geologic Hazards, Resources, and Environmental Planning*. Wadsworth Publishing Company.

Guillory, Dan. 1996. *When the Waters Recede*. Stormline Press.

Murck, Barbara W., Brian J. Skinner, and Stephen C. Porter. 1995. *Environmental Geology*. John Wiley & Sons.

Parrett, Charles, Nick B. Melcher, and Robert W. James, Jr. 1993. *Flood Discharges in the Upper Mississippi River Basin, 1993*. U.S. Geological Survey Circular 1120-A.

SAST. 1994. *Science for Floodplain Management into the 21st Century: A Blueprint For Change, Part V*. Preliminary Report of the Scientific Assessment and Strategy Team [SAST], Report of the Interagency Floodplain Management Review Committee to the Administration Floodplain Management Task Force, Washington, D.C.

St. Louis Post-Dispatch. 1993. *High and Mighty: The Flood of '93*. Andrews and McMeel.

U.S. Department of Commerce. 1994. *The Great Flood of 1993*. NOAA/NWS Natural Disaster Survey Report.

Tobin, Graham A. and Burrell E. Montz. 1994. *The Great Midwestern Floods of 1993*. Saunders College Publishing.

Wahl, Kenneth L., Kevin C. Vining, and Gregg J. Wiche, 1993. Precipitation in the Upper Mississippi River Basin, January 1 Through July 31, 1993. U.S. Geological Survey Circular 1120-B.

Ward, Roy. 1978. *Floods: A Geographical Perspective*. MacMillan Press.

Articles:

Boruff, Chester S. 1994. Impacts of the 1993 flood on Midwest agriculture. *Water International* 19: 212-215.

Bottorff, Harry. 1994. Rapid relief: GPS helps assess Mississippi River flood damage. *GPS World*: 23-26.

Braatz, Dean T. 1994. Hydrologic forecasting for the Great Flood of 1993. *Water International* 19: 190-198.

Brakenridge, G. Robert, James C. Knox, Earnest D. Paylor II, and Francis J. Magilligan. 1994. Radar remote sensing aids study of the Great Flood of 1993. *EOS Transactions* 75: 521-527.

Knapp, H. Vernon. 1994. Hydrologic trends in the upper Mississippi River basin. *Water International* 19: 199-206.

Kunkel, Kenneth E. 1994. A climatic perspective on the 1993 flooding rains in the Upper Mississippi River basin. *Water International* 19: 186-189.

Macilwain, Colin. 1993. Mississippi subsides, leaving debate in its wake. *Nature* 364: 747.

Macilwain, Colin. 1994. Nature, not levees, blamed for flood. *Nature* 369: 348.

Mauney, Thad and Harry R. Bottorff. 1993. FEMA assesses '93 flood disaster in Illinois using GPS and GIS technology. *ACSM Bulletin*: 25-29.

Petrie, G.M., G.E. Wukelic, C.S. Kimball, K.L. Steinmaus, and D.E. Beaver. 1994. Responsiveness of satellite remote sensing and image processing technologies for monitoring and evaluating 1993 Mississippi River flood developments using ERS-1 SAR, LANDSAT, and SPOT digital data., *Proceedings ASPRS/ACSM*: 176-182.

Pitlick, John. 1997. A regional perspective of the hydrology of the 1993 Mississippi River basin floods. *Annals of the Association of American Geographers* 87: 135-151.

White, W. Scott and Merrill K. Ridd. 1997. Visualization of a GIS-based water balance model. *Proceedings of the 18th International Cartographic Conference*: 2000-2007.

Case 3

Petroleum Geology: Persian Gulf vs. Overthrust Belt

Questions Explored:

How do the geologic histories of the Persian Gulf and the Overthrust Belt - Greater Green River Basin relate to their reserves of oil and natural gas?

What are the principal geologic similarities and differences between these two important oil-producing regions?

What are the major environmental impacts of hydrocarbon utilization?

How has US demand for Persian Gulf oil affected economic and cultural relationships with the region's petroleum producing countries?

Key Learning Outcomes:

The learner will be able to:

Locate the Persian Gulf and the Overthrust Belt on a world map.

Describe the major geologic processes that created oil and natural gas reservoirs in each region.

Compare and contrast the geologic formation, extraction process, economics, and environmental impacts of oil and natural gas exploitation in the two regions.

OVERVIEW

Location

The study areas examined here are located on opposite sides of the Earth. The **Persian Gulf**, currently the world's major producer of oil, is located in the Middle East, between 24° and 30° North latitude, and 48° and 56° East longitude. This shallow sea, connected to the Indian Ocean to the south, is surrounded by arid desert coastal plains and mountains. The Persian Gulf countries (Bahrain, Iran, Iraq, Kuwait, Qatar, Saudi Arabia, and the United Arab Emirates) contain more than two-thirds of the world's known oil reserves, and over one-quarter of the known natural gas reserves. Currently, the region produces about 4.3 billion barrels of oil per year. Around 250 of the region's known 300 oil deposits are in production. Sixty-nine of the oil fields are classified as **giants** (500 million to 5 billion barrels), 25 are **supergiants** (more than 5 billion barrels), at least 14 have known reserves greater than 10 billion barrels. The largest field, Ghawar in Saudi Arabia, has 83 billion barrels of oil.

Author:

Robert J. Moye and Sandra A. Zicus
Salt Lake City, Utah

Hydrocarbons have been known and used in this region throughout human history, beginning with the use of surface oil seeps and deposits of solid bitumens. The first major modern discovery took place in Iran in 1908, with more discoveries throughout the region between 1932 and 1938.

The **Overthrust Belt** and the adjacent **Greater Green River Basin** lie in the Rocky Mountains of the western United States, between about 41° and 44° North and 110° and 112° West latitude. Covering portions of Utah, Wyoming, and Idaho, they are regionally important producers of oil and natural gas, but do not contribute significantly to world production. Petroleum, in the form of "tar" or "oil springs," has been known in southwestern Wyoming since the early 1800s, and was used by westward-moving settlers to grease the wheels of their wagons. The first subsurface oil discovery in the Overthrust Belt was made by the Union Pacific Railroad Company while drilling for water in 1900. Significant discoveries were made in 1924 at the transition zone between the Overthrust Belt and the Green River Basin, but exploration was intermittent and most discoveries shallow until after World War II. Intensive modern exploration and development did not begin until 1975, with the discovery of oil and gas at Pineview Field in Utah. To date, twenty-nine oil and natural gas fields have been discovered in the Overthrust Belt, with twenty-four in production in 1995. The Greater Green River Basin to the east is primarily a producer of natural gas, with numerous producing fields scattered over a large area of southwestern Wyoming.

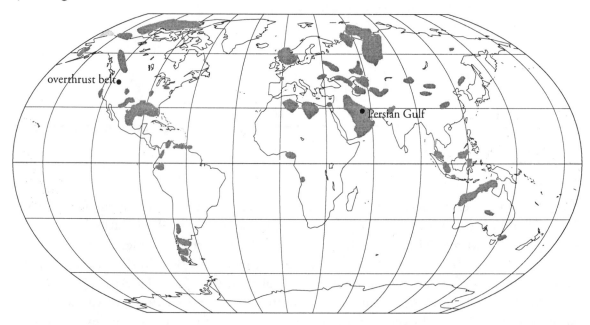

Figure 1. World map showing areas of major known oil and natural gas reserves with the location of the Persian Gulf and Overthrust Belt highlighted.

Primary Objectives
The primary focus of this case study is to foster an understanding of the origin and distribution of oil and gas in two very different geologic environments, and to explore the social, economic, and environmental issues associated with hydrocarbon exploration and production in both regions.

How do the geologic histories of the two regions relate to their reserves of oil and natural gas?

The Persian Gulf

Organic-rich marine sediment host rocks were deposited in shallow tropical seas on a stable continental shelf in the Persian Gulf area throughout much of the Paleozoic and Mesozoic Eras. The earliest sediments include a thick basal sequence of 800 to 500 million year old **evaporites**. During subsequent history, **salt domes** rose from these evaporites through the overlying sediments, creating traps for oil and gas accumulation. Sedimentation was interrupted by a mountain-building event known as the **Hercynian Orogeny** (about 350 million years ago). Deposition of shelf sediments resumed about 300 million years ago during the Late Carboniferous and continued for the next 100 million years through the Late Permian to Early Triassic, when the area subsided during the formation of the Tethys Sea to the north and a thick sequence of limestones was deposited. Burial and maturation of the pre-Carboniferous source rocks by these later strata produced oil and gas that accumulated in Permian limestone reservoirs and in traps formed by salt domes that rose from the basal evaporites at this time.

Throughout the Mesozoic, the Persian Gulf region remained a shallow marine shelf area adjacent to the Tethys Sea with periodic tectonic movements, climatic changes, and variations in sea level affecting the composition and thickness of sedimentation. A deep trough formed in the Late Jurassic (about 160 million years ago), where very organic-rich bituminous shales were deposited and later buried beneath evaporites and carbonates. This sequence became a major source for subsequent hydrocarbon development in the area. Sandstone deposition alternated with the formation of shale basins throughout the Cretaceous, forming a further close association of source rocks and potential reservoirs.

In the Late Cretaceous (about 90 million years ago), with the onset of the Alpine Orogeny, the Tethys Sea began to close as the African and Eurasian tectonic plates moved together. Older continental rocks to the northeast were thrust southwest along the **Zagros Fault Zone**. Adjacent Gulf area sediments were strongly folded, forming the Zagros Mountains of western Iran. These **anticlines** formed traps for subsequent oil and gas migration. A deep marine trough known as the Zagros Basin, a southern arm of the Tethys Sea, developed southwest of the mountains along the present trace of the Persian Gulf, with a shallow carbonate shelf to the west on the Arabian Peninsula. Sediments deposited in this trough over the next 70 million years formed important Tertiary source rocks and reservoirs.

Tectonic activity in the Persian Gulf area over the last 20 million years has been related to the opening of the Gulf of Aden and the Red Sea to the west, which are slowly pushing the Arabian Peninsula to the northeast. Relative tectonic quiescence during this time has allowed previously formed oil in the basin sequence to migrate upward into new reservoirs.

Oil and natural gas fields in **Saudi Arabia**, west of the Persian Gulf, occur within shelf sediments and are often associated with traps formed by salt domes that have risen through the overlying sediments from the basal evaporite sequence. These domes probably began to rise during the Permian, and reservoir rocks are often of Late Permian-Early Triassic age. Late Cretaceous tectonism initiated a second phase of salt movement below the Zagros Basin in **western Iran, Kuwait,** and **Iraq.** Here, some salt domes reached the surface and continued to be active until the Quaternary. Associated reservoirs are of Jurassic and Cretaceous age, with oil derived from Upper Jurassic to Lower Cretaceous source rocks. Anticlines formed by folding in the Zagros Mountains of Iran during the Oligocene to Miocene (20 to 30 million years ago) are the dominant reservoirs east of the Persian Gulf, and contain oil and gas that have migrated from older source rocks.

The Overthrust Belt and Greater Green River Basin

The Overthrust Belt is the eroded remains of a large mountain range created during the Sevier Orogeny (120 to 50 million years ago). It is an easterly bulge, or **salient**, of the greater **Cordilleran Thrust Belt** that extends for over 5000 miles along the western margin of North America from Alaska to Mexico. There were major changes in the convergence of the North American and Pacific tectonic plates during this time. A series of volcanic island arcs and other accreted terranes collided with the western margin of North America, buckling it and pushing thick sequences of sedimentary rocks westward in a series of stacked thrust faults. In the Overthrust Belt, these thrusts piled up near the present Utah-Wyoming border to form the Sevier Mountains, stretching through present-day Idaho, Wyoming, and Utah.

Prior to this, the western limit of North America was in central Nevada. From the Late Proterozoic through the end of the Paleozoic, the Utah-Wyoming area was a shallow continental margin located near the equator. Thousands of feet of limestones and sandstones were deposited along this margin. During the Triassic and Jurassic, the assembly of the supercontinent Pangea brought major tectonic and climatic changes to the area, and mudstones and sandstones were deposited along shorelines and on the broad flood plains of rivers.

Throughout much of the Cretaceous, a broad, shallow seaway extended across central North America from the Gulf of Mexico to Alaska west of the growing Sevier Mountains. Organic-rich marine sediments deposited in this seaway were buried beneath younger sediments that were deposited ahead of the advancing thrusts.

Burial beneath thrust faults of the Sevier Mountains during the early Tertiary resulted in maturation of the organic matter in the Cretaceous marine sediments. During the following long period of tectonic quiet, the oil and gas have migrated upward into numerous reservoirs that were formed by folding and faulting in the rocks above the thrust faults. The most common traps are found in the Jurassic Nugget Sandstone in the crests of anticlines. A thick sequence of sedimentary rocks, including potential hydrocarbon source rocks, are also present below the Cretaceous marine sediments, but they are so deeply buried that they are outside the **petroleum window** and do not produce oil or gas. Oil and gas reservoirs occur only along the eastern side of the

Overthrust Belt, as many reservoirs to the west have been breached by erosion and the underlying potential source rocks no longer yield hydrocarbons.

The Greater Green River Basin is composed of a thick sequence of sediments deposited in a large, interconnected series of fresh water lakes that formed along the eastern margin of the Overthrust Belt during the Eocene Period, about 40 million years ago. Fossils of alligators and palm fronds show that the climate was subtropical and biologic productivity high. Because the lakes had interior drainage, a great thickness of organic-rich sediments accumulated, with sequences of **oil shale**, and the world's largest deposits of **trona**, **sodium carbonate** and **bicarbonate salts**. The basin existed for four to eight million years, and developed on top of the thick regional sequence of Cambrian to Cretaceous age sediments, including some organic-rich **shales** and **limestones**. Burial and maturation of some of these older source rocks formed large volumes of natural gas that have migrated into anticlinal and stratigraphic traps formed within the overlying Green River Basin sediments. The sediments of the Green River Basin itself were never within the petroleum window and have not yielded oil or gas; however, research has been conducted on the potential for extraction of oil from the oil shales.

What are the principal geologic similarities and differences between these two important oil-producing regions?

The geologic history of both the Persian Gulf and the Overthrust Belt share the critical events necessary for the formation and accumulation of recoverable oil and natural gas:

1. deposition and preservation of potential source rocks
2. burial and maturation of source rocks to produce hydrocarbons
3. migration and accumulation of oil and gas in suitable reservoirs

However, the number, duration, and geographic extent of oil-forming events have differed significantly between the two regions.

The **Persian Gulf** region has remained at low latitudes, within 30° of the Equator where biologic productivity is high, for the last 250 million years. The region has been the site of repeated accumulation of organic-rich marine sediment source rocks in large depositional basins, with subsequent formation of suitable hydrocarbon traps. Oil and natural gas are derived from source rocks ranging in age from 500 to 150 million years that are present over a large geographic area, and have accumulated in reservoirs formed by stratigraphic, salt dome, and anticlinal traps ranging in age from 250 to 20 million years.

The geologic history of the **Overthrust Belt** area is similar to that of the Persian Gulf area. However, only one significant source of oil and gas is present within the petroleum window, and only one event produced suitable traps. The limestones and shales of the Proterozoic through Permian marine shelf include potential oil source rocks, but are outside the petroleum window in the Overthrust Belt area. The Triassic and Jurassic strata are important as reservoirs, but are not significant source rocks. The only places where the source rocks of the Cretaceous Seaway are both within the petroleum window and adjacent to structural traps are along the eastern side of the Overthrust Belt and in the adjacent Green River Basin.

Some of the principal geological factors that affect the economics of exploration, discovery, and production of oil and gas include the size and complexity of the reservoirs and their depth from the surface. Many of the oil and gas fields in the Persian Gulf area are large, of simple structure, and located at relatively shallow depths. By contrast, the traps in the Overthrust Belt - Green River Basin are relatively small by Persian Gulf standards, very complex in structure, and therefore harder to find and exploit.

THE HUMAN DIMENSION

What are some of the major environmental impacts of hydrocarbon utilization?

The potential environmental hazards and impact of hydrocarbon utilization increase from the exploration phase to production, to refining, to consumption. Impacts from **exploration** for oil and natural gas include surface disturbance from drilling and road building, and the release of hydrocarbons through leakage and **blowouts**, the explosive venting of oil or gas from a high pressure reservoir when it is penetrated by a drill hole. Blowouts are relatively rare, but can result in more oil spilled than from a tanker accident. Oil and gas **production** results in significantly greater surface disturbance as as the local infrastructure to support the operation is developed. Often, surface disturbance is greatly magnified as additional wells are drilled into the reservoir, and as the pumps, pipelines, and storage tanks are constructed in order to maintain th new wells. Oil is often associated with water, sometimes saline, that must be disposed of. Today, this water is often reinjected into the reservoir to enhance oil recovery. **Subsidence**, or sinking of the surface as oil is withdrawn from reservoirs below, was a problem in the past, but the modern practice of injecting water prevents collapse. Many different gases may accumulate in both oil and natural gas reservoirs, including carbon dioxide (CO_2), nitrogen (N_2), helium (He), argon (Ar), and hydrogen sulfide (H_2S). Hydrogen sulfide is odorless, colorless, and potentially lethal. Oil and natural gas fields that produce H_2S are constantly monitored for leaks, and the gas is collected for disposal or sale. Natural gas associated with oil reservoirs was often burned off or **flared** in the early days of the industry, contributing to atmospheric pollution. Only about 1% of natural gas is now flared in North America, and about 8.5% in the Middle East.

The greatest environmental hazards are associated with the **transport, refining, and utilization** of hydrocarbon products. Transport of oil or gas from the field to the refinery, from refinery to distribution centers, and from distribution to market by pipeline, truck, train, or boat usually involves at least minor loss through leakage, and carries the potential for major spills. About 6,000 oil tankers traverse the Persian Gulf, and some estimates indicate that about 40% of these tankers leak, spilling the equivalent of 25,000 barrels of oil per day into the Gulf.

The refining process invariably results in frequent minor leakage and occasional serious spills. Most refinery sites are significantly polluted, potentially threatening local groundwater. The distillation process produces some non-marketable gases, and these are flared at some refineries, contributing to air pollution. Refineries remove sulfur from oil and natural gas to meet air quality standards, and now account for 55% of US production of sulfur.

Global CO_2 emissions tripled between 1950 and 1990, to about 8.5 billion tonnes annually. About 27% of this total is from the burning of oil products and 11% from natural gas, another 27% from burning coal, 2% from cement production, and 33% from the clearing and burning of forests. Increased atmospheric levels of CO_2 are associated with acid rain and greenhouse warming. Although there is disagreement as to the degree and the specific effects of global warming due to the build-up of greenhouse gases, most scientists agree that warming is likely and that CO_2 emissions are a significant factor. Of growing concern among health organizations is the hazard from **particulates**, fine solid particles or liquid droplets of hydrocarbons produced by the combustion of petroleum products, especially oil and gasoline. Automobile and truck exhaust are the major source, followed by some industrial and power plants. These particles are inhaled, and can damage the respiratory system.

The United States depends on oil products for 97% of its energy for transportation, accounting for about two-thirds of the total oil used in the country. If Americans continue the recent shift to less fuel-economic vehicles such as sport-utility vehicles, minivans, and pickup trucks, and keep driving around 3% more miles each year, the environmental impacts of oil production and use will remain a serious problem.

How has US demand for Persian Gulf oil affected economic and cultural relationships with the region's petroleum-producing countries?

In 1996, 48% of the oil consumed in the United States was imported at a cost of around $500 million dollars. Almost 20% of all oil imported into the United States came from the Persian Gulf region, and it is estimated that this percentage will rise to more than 25% by the year 2000. Clearly, this dependence on imported oil, especially from the Middle East, has had a major effect on US foreign policy and foreign relations, and will continue to do so.

The oil resources of the Persian Gulf area began to achieve prominence at the end of World War I. At that time, the Ottoman Empire was destroyed and European allies, especially Great Britain and France, began to carve the area up into new nations and dynasties, beginning the global power struggle to gain control of Middle Eastern oil. For the first decade after the end of the war, Great Britain was a major player. Their major concern at this time was to maintain access to known oil and reported reserves in Mesopotamia.

A 1932 discovery of oil in Bahrain by the Bahrain Petroleum Company signaled a change in focus. Not only did this bring American oil interests into the region, it also caused geologists to begin examining geologically similar structures on the Arabian mainland. The following year, Standard Oil of California signed a contract with Saudi Arabia to begin petroleum exploitation in that country. The Gulf Oil Corporation of the United States and British Petroleum Company, working with the government of Kuwait, joined forces to create the Kuwait Oil Company. The race was on.

During World War II, denying the oil reserves of the Middle East to the Third Reich was critical to the defeat of Germany. The German campaign to seize control of the region was resisted by the Allies, and finally stopped at El-Alamein in 1942. In the same year, the United States Army set up the Persian Gulf command in Iran in order to send oil supplies to the Soviet Union for their continued fight against Germany. These wartime demands for oil began the shift from dependence on the United States (which then produced more than 60% of the world's oil) to the Middle East.

After the war, nationalist sentiments increased among the Persian Gulf states. In 1951, the Iranian government nationalized the oldest Western oil concession, the Anglo-Persian Oil Company. This threatened American interests and, in 1954, the government of Iran was overthrown with assistance from the United States. This event ultimately led to the Iranian Revolution, the overthrow of the Shah, and much of the present hostility between the United States and Iran. A new consortium of oil companies, dominated by the Americans, took over the oil concession.

As nationalist attempts and counteraction by Western nations continued throughout the 1950s and 1960s, it became clear that control of oil resources was the dominant factor. In September of 1960, the Arab states created the Organization of Petroleum Exporting Countries (OPEC) in an attempt to maintain control of their resources. More nations joined OPEC and nationalized their oil industries, resulting finally in the OPEC oil embargo to the West during the Arab-Israeli War in 1973. Oil prices jumped, and the United States increased government involvement in the Middle East, encouraging the development of military power in Saudi Arabia and Iran.

The United States military has remained heavily involved in the region. In 1980, President Carter announced that "an attempt by any outside force to gain control of the Persian Gulf region will be regarded as an assault on the vital interests of the United States of America, and such an assault will be repelled by any means necessary, including military force." During the Iran-Iraq War in 1981, President Reagan made the message even clearer when he declared that the United States would not "stand by and see that taken by anyone that would shut off that oil." During the first Reagan administration, the US spent almost $1 billion on military construction in the Persian Gulf region.

After the end of the Iran-Iraq War in 1988, Iraq had accumulated a debt of about $70 billion dollars. Meanwhile, the tiny country of Kuwait, with 10% of the world's oil reserves, raised their production to almost 1.9 billion barrels a day, leading to a sharp decline in oil prices. This action, combined with a long-standing border dispute between Iraq and Kuwait, led to the Iraqi invasion of Kuwait on August 2, 1990. Oil exports from Kuwait to the West ceased and within two months, world oil prices more than doubled. Other OPEC nations, especially Saudi Arabia, increased production to meet world demand.

The United Nations, urged by the United States, responded by sending troops into Saudi Arabia, and on January 28, 1991 the Iraqi army set fire to more than 700 Kuwaiti oil wells and began releasing crude oil into the Persian Gulf. UN troops attacked on February 24, and Iraq surrendered five days later. The tremendous environmental, economic, and human costs of this short war are still being calculated and will continue to be paid for generations. As long as world dependence on oil continues, so will conflicts in the Persian Gulf.

CONCEPTUAL QUESTIONS

What is the role of plate tectonics in the formation of oil and natural gas reserves?

What geologic factors should be considered when evaluating the potential of an area for the presence of oil and natural gas? What about political and environmental factors?

Are fossil fuels renewable resources? Is dependence on fossil fuels as a primary source of energy sustainable in the long-term? Why or why not?

What are some possible ways of decreasing the dependence of the United States on fossil fuels without increasing other negative environmental, social, or political impacts?

SOME RECOMMENDED READINGS

Books:

Christensen, J. W. 1996. *Global Science.* Kendall/Hunt Publishing Company, Dubuque, Iowa.

Cocks, L. R. M. ed. 1981. *The Evolving Earth.* British Museum of Natural History. Cambridge University Press, Cambridge, UK.

Coogan, James C., and Frank Royse, Jr. 1990. *Overview of recent developments in Thrust Belt interpretation.* in Roberts, Sheila. ed. *Geologic field tours of western Wyoming and parts of adjacent Idaho, Montana, and Utah.* Geological Survey of Wyoming, Public Information Circular No. 29, p. 88-124.

Craig, J. R., D. J. Vaughan, and B. J. Skinner. 1988. *Resources of the Earth - Origin, Use, and Environmental Impact,* 2nd Ed., Prentice Hall, NJ.

Hyne, N. J. 1995. *Nontechnical Guide to Petroleum Geology, Exploration, Drilling and Production.* Pennwell Publishing Company, Tulsa, OK.

Kesler, S. E. 1994. *Mineral Resources, Economics and the Environment.* Macmillian College Publishing Company, New York.

Shannon, P. M., and D. Naylor. 1989. *Petroleum Basin Studies,* Grahan & Trotman, London, UK.

Snoke, A. W., J. R. Steidtmann, and S. M. Roberts, eds. 1993. *Geology of Wyoming*, Memoir No. 5, Geological Survey of Wyoming, Laramie, WY.

Ver Ploeg, A. J. and R. H. De Bruin. 1982. *The Search for oil and gas in the Idaho - Wyoming - Utah Salient of the Overthrust Belt.* Report of Investigations No. 21. Geological Survey of Wyoming. Laramie, WY.

Articles:

Al-laboun, A. A. 1988. The distribution of Carboniferous-Permian siliciclastic rocks in the greater Arabian Basin. *Geological Society of America Bulletin.* vol. 100. p. 362-373.

DeCelles, Peter G. and Gautam Mitra. 1995. History of the Sevier orogenic wedge in terms of critical taper models, northeast Utah and southwest Wyoming. *GSA Bulletin,* v. 107, no. 4, p. 454-462.

Gealey, W. K. 1977. Ophiolite obduction and geologic evolution of the Oman Mountains and adjacent areas. *Geological Society of America Bulletin.* vol. 88. p. 1183-1191.

Case 4

Using GIS and Remote Sensing to Monitor Urban Habitat Change

Questions Explored:

What is the unique situation of the Salt Lake Valley in regard to land use and land cover change in an urban-to-rural transition zone?

What are the biophysical characteristics of a city in a transect from core to periphery?

What are the "new tools" of Earth Systems Science?

How are these tools used to monitor rural-to-urban land use and land cover conversion?

What implications does urbanization have on global environmental change?

Key Learning Outcomes:

The learner will be able to:

Understand the principles of human population change within an urban environment.

Understand the general impacts that human population growth and population concentration have on the physical geography of a city.

Understand the fundamental principles of remote sensing of earth surface phenomena.

Define the characteristics of land use and land cover within an urban environment.

Comprehend the contributions of urban areas toward global environmental change.

OVERVIEW

Location

This case study takes place in the Salt Lake Valley of Utah, which comprises most of Salt Lake County and is the major population center for the state. The study area considered here is the eastern suburbs of the Salt Lake Valley, including portions of Salt Lake City, Draper, Murray, Sandy, and Midvale, Utah. Although we allude to the entire Salt Lake Valley, the focus area of this study comprises an area running north to south from central Salt Lake City, Utah, along the eastern benches (foothills) of the Wasatch Front to Draper, Utah. The research area is illustrated in Figure 1. The locator map shows the study area location in relation to the state of Utah.

The valley has a unique environmental situation — bounded to the north and east by the Wasatch Mountains, to the west and south by the Oquirrh Mountains, and to the north-west by the Great Salt Lake (discussed in greater detail in case study number 6).

Author:

James D. Hipple
University of Missouri-Columbia

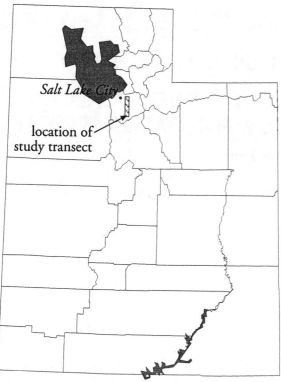

Figure 1. The Salt Lake County research study area and its location within the State of Utah.

The mountainous terrain potentially limits human population growth and urban sprawl by putting strict bounds on the amount of developable land, resulting in greater population density. Surrounding the Great Salt Lake are a number of interesting, but fragile, ecosystems dependent upon the unique interplay between the geosphere, biosphere, and hydrosphere.

Primary Objectives

This case uses **land use** and **land cover change** as the framework for exploring urban habitat change, focusing on the fundamentals of geographical information systems and remote sensing techniques for monitoring and modeling such change in a rural-to-urban transition zone. **Land use** refers to what a particular area is being used for, such as urban residential, commercial, or agricultural, and **land cover** refers to the physical ground cover, such as forest, corn crop, asphalt, or grass.

In this study, an emphasis is placed on monitoring the biological, ecological, and human implications of urban growth in a suburban environment. Geological and geomorphological examples and implications are presented on the *GeoSystems Today: an Interactive Casebook* web site.

BACKGROUND

What is the unique situation of the Salt Lake Valley in regard to land use and land cover change in an urban-to-rural transition zone?

The Salt Lake Valley is a unique place in which to study the evolution of the urban environment. Prior to 1847 and the influx of the Mormon Pioneers, the landscape consisted of a mixed sagebrush community on the valley floor and limited riparian communities along a number of small creeks on the east side of the valley and along the centrally located Jordan River. In addition, there was a shrubby foothill community, dominated by Gambel's oak (*Quercus gambelii*).

Today, if you were dropped down in the middle of the Salt Lake metropolitan area you probably would not be able to distinguish it from a typical eastern United States city (except for the looming mountains!). Before extensive human settlement, the eastern United States was dominated by established climax forests. After settlement, the forests were partially cleared for farming, villages, and cities. Remnants of the forest were maintained over time as single trees or clumps of trees. Occasional old growth or secondary growth trees and their associated plant and animal communities can still be found in these eastern cities. The Salt Lake Valley – in contrast to the eastern city — was essentially a pristine semi-arid valley. Upon their arrival, the Mormon immigrants started a pattern of landscape alteration which mimicked the environment from which they came, not only in environmental character, but in architectural style, culture, and other practices, as well. The result is a similar pattern of tree and other plant species within both urban environments, in the east from natural forest habitat destruction, and in the Salt Lake Valley from urban forest habitat creation.

Human population dynamics (inmigration, births, deaths, and outmigration) has played an important role in reshaping the environmental character of the Salt Lake Valley. The increase in suburban populations and decrease in the population of the central city is characteristic of trends observed from the 1960s through the 1980s observed nationwide (Figure 2). Many land use conflicts surfaced as a result of these migration trends. A nationwide urban to rural population shift in the 1970s and an increase in the degree of economic prosperity introduced many new problems into rural areas and urban hinterlands. During the 1980s a new trend emerged, a reversal, where metropolitan county population growth out-paced that of non-metropolitan counties. These growing counties, often located on the periphery of the central city, experienced many land-use conflicts between rural and urban oriented land uses.

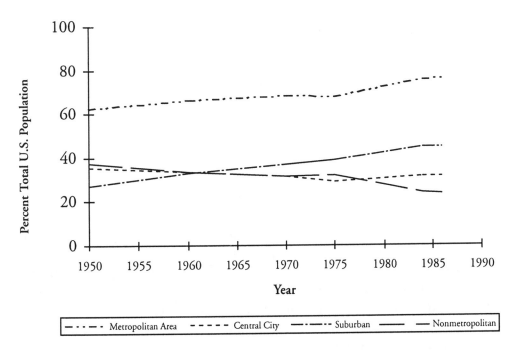

Figure 2. Percent total US population by residential category.

The rural population boom of the 1970s and increasing suburbanization of the 1980s is characteristic of much of the United States, including Salt Lake County. Throughout much of the rural-urban fringe of Salt Lake City many "rural land use conflicts" are present, including the loss of farmland and open space, and the presence of waste

41

dumps, small and large scale industrial plants, and intensive residential and commercial uses. According to the US Census Bureau, the population within Salt Lake County has increased by more than 250 percent since 1950, while the population within Salt Lake City has remained relatively static. This statistic tells a great deal about events which have occurred within the county during the past 50 years — a pattern observed in most American cities of similar size. The central, or historic city, has experienced stagnating growth or even population decline, whereas the suburbs have seen blossoming rates of growth. Figure 3 illustrates this relationship graphically. The line graph plots population over time for Salt Lake County and Salt Lake City between the years 1950 and 1990. From this illustration we can envision the rapid development in residential suburbs and visualize the great pressures being placed on the natural resources of the state. The density of development in Salt Lake County, by far not the densest nationwide, is characterized by sprawling suburbanization rivaling that of Los Angeles, California. The prosperity of the 1990s has spawned a great deal of suburbanization where lot sizes of one half-acre to one acre are the norm. Needless to say, this is inefficient use of land and greatly contributes to the disappearance of open space and farmland. The increase in development puts great pressures on the environment, affecting water quality, wildlife, and even the quality of life for the human residents within the valley.

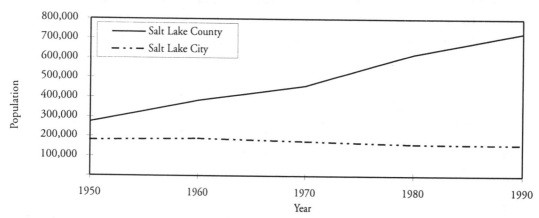

Figure 3. Salt Lake County population trend from 1950 to 1990 (Source: US Census Bureau).

There are many problems associated with changes in land use and land cover, particularly in the area referred to as the rural-urban fringe (RUF). Problems here are typically more intense than those occurring in other areas because the RUF is a dynamic area with an intermingling of rural and urban land uses. This intermingling of uses leads to a number of problems — land use change being one of the largest. Conflicts associated with rural-urban fringe development include the conversion of prime agricultural land to urban-oriented uses, increasing population and associated social problems, and environmental degradation.

What are the biophysical characteristics of a city in a transect from core to periphery?

Scientists often use models to help understand and explain difficult and complex problems. Models can be complex mathematical equations or a simple icon representing the relationships or objects being studied. One thing all models have in common is that they are idealized structured representations of reality. In other words, they are generalizations of the real world that leave in the important elements and disregard or control for the peripheral elements. Generating a model is one method for understanding the complex dynamics of the urban biophysical (**biological** and **physical**) environment. A number of urban models have been developed using important **biophysical** measures such as **biomass**, **plant species density** and **diversity**, and **net primary productivity (NPP)**. These models group land uses and land covers into relatively homogeneous units. In assessing the ecological pattern of a city, there is a direct correspondence between social and economic variability within a region and the region's biophysical characteristics and resulting land cover. One very descriptive model is the Biological Zone Model illustrated in Figure 4. Robert Dorney, a researcher from the University of Toronto, has identified nine principal biologic zones which emanate from the center of a city in a concentric ring pattern.

Urban Biological Zones (Dorney 1979)	Urban RS Typology (Ridd et al. 1983)	Net Primary Productivity (NPP)	Biomass	Plant Diversity	Plant Types
Cliff/Organic Detritus (Central Business District--CBD)		minimal	low	low	weeds and exotic flowers
Old Urban Savannah (housing 50 years plus)	Residential Mature R_m	low	medium to high	medium	dominated by native and exotic plants
Intermediate Urban Savannah (housing 15-50 years old)	Residential Established R_e	high P>R	medium	high	dominated by native and exotic plants
New Urban Savannah (housing 0-15 years old)	Residential New R_n	low P>R	low	low	exotics dominate; particularly various sp. of grasses
Abiotic Weedy Complex (construction zone)	Residential Incipient R_i	low	low	high	native and exotic (dominated by 'weeds')
Remnant Ecosystems (i.e., woodlots and marshes)		stable P=R	high	high	numerous native plants often with some exotics encroaching
Agricultural Ecosystem (cropland)	Agricultural A_g	high P>R	low	low	cultivars (often a monoculture)

urban core (at left, top) ... *periphery* (at left, bottom)

Figure 4. Biological zonation within the urban environment (modified from Dorney 1979 and Ridd et al. 1983).

Starting at the city periphery — close to the agricultural hinterland — we find either the **Remnant Ecosystem/Natural Island** zone, consisting of patches of undisturbed or slightly disturbed natural areas, or an **Agricultural Ecosystem**, often composed of monocrops and rangeland. The **Abiotic/Weedy Complex** is the zone of new housing construction, which at one time was either a Remnant Ecosystem/Natural Island or an

Agricultural Ecosystem. The natural or crop vegetation in the Abiotic/Weedy Complex is removed almost completely before construction begins.

Over time, the Abiotic/Weedy Complex is transformed into the **New Urban/ Savannah**, a residential housing zone which is characterized by three to thirty percent crown canopy cover, low biomass (evident from the lack of large trees and predominance of grass as a ground covering), and low avian diversity. This includes residential areas fifteen years of age or less.

As the housing ages, new units are built at the periphery of the city, resulting in a new wave of urbanization. The Abiotic/Weedy Complex is now found further away from the core of the city, and there is also a new New/Urban Savannah. Residential areas that were built 15 years ago start to take on the characteristics of the **Intermediate Urban/Savannah**, having twenty to forty percent crown cover, rapidly increasing biomass (illustrated by high net primary productivity (NPP) where photosynthesis outpaces respiration), and the highest avian diversity. This zone is typical of residential areas between fifteen and fifty years of age.

Closer to the center of the city, we find the **Old Urban/Savannah**. This area is characterized by having forty percent or greater crown canopy, decreasing biomass, and intermediate avian diversity. At the center of the city — the heart of the city — we find the **Cliff/Organic Detritus Zone**, named for its almost complete lack of vegetation and the cliff-like environment created by tall buildings towering over the canyon-like streets.

The model derived from Dorney's work is one method for understanding the complex biophysical characteristics of the urban environment. We must keep in mind that the zones change as the city ages and expands at the periphery. Some zones become larger (like the Old/Urban Savannah in a city that is over 100 years old), and others, like the Abiotic/Weedy Complex construction zone, keep migrating to the edge of the city.

What are the "new tools" of Earth Systems Science?

The "new tools" of **Earth Systems Science (ESS)** include tools such as **geographic information systems (GIS)** and **remote sensing (RS)**. Other tools include computer visualization, as well as **spatial and aspatial modeling**. Although they may be considered "new" in many disciplines, these tools and the techniques that they employ to analyze spatially-referenced biophysical data have been around for many decades in geography, geology, and other ESS fields. What is "new" about these tools is their wide use today as compared to the past.

A **GIS**, or **geographic information system**, is a computer program designed to input, store, and analyze **spatially-referenced data**. Spatially-referenced data are pieces of information that have some sort of location attribute associated with them. The location attribute is usually some geographic reference system (coordinate system) such as longitude-latitude, state-plane coordinates, or Universal Transverse Mercator (UTM) northing and easting. The GIS acts as a large spatial database which an analyst can use to perform simple non-spatial query functions (such as tabulating the population of a state by summing the populations of its counties), or to answer questions having some spatial component (such as determining the population within a five mile radius of a particular elementary school).

The term **remote sensing** refers to the practice of deriving information about the Earth's surface from an overhead perspective, and includes aerial and satellite photography as well as **electromagnetic imaging**. In electromagnetic imaging, airborne sensors (mounted on a plane), or space-based sensors (on a satellite) collect data about the physical landscape as the instrument flies overhead and records the electromagnetic radiation **emitted** or **reflected** from the surface of the Earth. The data are commonly collected in "bands" based on wavelength, and data from different bands are combined to create "false color" images – i.e. colors are assigned arbitrarily to the images. In order to interpret the images, you need to have some knowledge about the physical objects of a landscape, including features such as vegetation, soil, water and buildings.

Once the photographs or digital imagery have been acquired, transformations can be made to reveal specific kinds of information. What transformations you do to the imagery depend upon what type of information you want. A single photograph or image can yield information about geology, soils, hydrology, vegetation, or even land use. Data obtained from remote sensing are often combined with ancillary data within a GIS.

How are these tools used to monitor rural-to-urban land use and land cover conversion?

The usefulness of these tools in monitoring rural-to-urban land use and land cover conversion depends upon two factors: the scale of observation (**spatial resolution**) and the frequency of observation (**temporal resolution**). The spatial resolution (the fineness of detail recorded by the satellite or airborne sensor) determines the types of objects we can observe. A sensor with relatively coarse resolution (80m to 1.1 km or larger) will often not detect features such as buildings or roads. A sensor with relatively fine resolution (ranging from 1m to 30m) will pick up these features to varying degrees. The temporal resolution of these sensors (how often data are collected over a given location), varies from daily to monthly, and can be even more infrequent. Seasonality and weather all take a toll on acquiring data when you need it.

Once we find the sensor with the proper spatial and temporal resolution, we can construct a land use and land cover map. Remote sensing data are collected, and a number of land surface covers are identified by means of statistical and spatial analysis procedures. Land use and land cover maps for multiple time periods can help researchers, city planners, and citizens understand where the greatest amount of rural-to-urban land use conversion is taking place. This understanding helps them determine where their energies should be focused in order to provide city services to encourage growth, or to enact zoning regulations to preserve open space and farmland, and regulate growth.

What are the implications for the human dimensions of global environmental change?

Only recently has there been recognition of urban systems as being globally significant, both in terms of their function as human habitats and their effect on Earth's natural systems. There is an increasing interest in treating the urban phenomenon as an important ecological entity. Researchers are becoming aware that to understand the impacts of humans on global environmental change they need to look where people are concentrated — in the city!

Statistics regarding the concentrations of human populations in cities are frightening. More than 50% of the Earth's population lives in urban places, and the percentage is rising on every continent. The number of extremely large cities (8 million people or more) is also growing as a result of global population increase. These cities, known as "mega-cities", are occurring not only in developed nations but also, at an even faster rate and in greater number, in undeveloped and developing nations. Many of these cities are experiencing population growth that is 3 times faster than the global average. It is estimated that by the year 2025 there will be 31 mega-cities. This is dramatic considering that in 1965 there were only two mega-cities (New York and London). These mega-cities will dominate the flow of regional and global resources, cover over a potentially significant share of the planet's best soil, and generate huge quantities of waste material, as well as profoundly influence their local biogeophysical environments. As such, they will constitute a kind of "organism" with its own metabolism and environment-altering characteristics.

The importance of this topic is frequently addressed in the rural-urban fringe of Salt Lake City, Utah, as well as across the United States. The best way to gain understanding of local impacts is to visit the site, identify the specific characteristics, and determine the local attitude to the land use - land cover conversion process. To some people, urban sprawl and land use conversion are signs of progress, an expanding economy, and prosperity. To others, sprawl is a blight on the landscape, an effect of uncontrolled growth that cannot be sustained. It is essential for developers, farmers, elected officials, and the resident population to address the issues brought forth, because land use change impacts the quality of life for all concerned.

CONCEPTUAL QUESTIONS

What particular areas of your city are experiencing rural-to-urban land use conversion?

What parts of the country are experiencing rapid growth? Which are in ecologically sensitive areas?

Are the impacts of rural-to-urban land use conversion dramatically different in the developing world versus the developed world?

SOME RECOMMENDED READINGS

Books:

Adams, Lowell W. 1994. *Urban Wildlife Habitats: A Landscape Perspective.* Minneapolis: University of Minnesota Press.

Douglas, Ian. 1984. *The Urban Environment.* Baltimore, MD: Edward Arnold.

Gilbert, O. L. 1989. *The Ecology of Urban Habitats.* New York: Chapman & Hall.

Spirn, Anne W. 1984. *The Granite Garden: Urban Nature and Human Design.* New York, NY: Harper Collins Publishing.

Whitehand, J.W.R. 1991. *The Making of the Urban Landscape.* Cambridge, MA: Blackwell Publishers.

World Resources Institute. 1996. *World Resources 1996-97: A Guide to the Global Environment (The Urban Environment).* Washington, D. C.: World Resources Institute.

Articles:

Dorney, Robert S. 1979. The ecology and management of disturbed urban land. *Landscape Architecture.* 69 (3):268-272.

Dagg, A. I. 1970. Wildlife in an urban area. *Canadian Naturalist.* 97:201-212.

Imhoff M. L., W. T. Lawrence, D. C. Stutzer, and C. D. Elvidge. 1997. A technique for using composite DMSP/OLS City Lights satellite data to map urban areas. *Remote Sensing of Environment.* 61(3):361-370.

Imhoff M. L., W. T. Lawrence, C. D. Elvidge, T. Paul, E. Levine, and M. V. Privalsky. 1997. Using nighttime DMSP/OLS images of city lights to estimate the impact of urban land use on soil resources in the United States. *Remote Sensing of Environment.* 59(1):105-117.

McDonnell, M. J., and S. T. A. Pickett. 1990. Ecosystem structure and function along urban-rural gradients: an unexploited opportunity for ecology. *Ecology* 71 (4):1232-1237.

Ridd, M. K. 1995. Exploring a V-I-S (vegetation-impervious surface-soil) model for urban ecosystem analysis through remote sensing: comparative anatomy for cities. *International Journal for Remote Sensing* 16(12):2165-2185.

Soule, M. E. 1991. Land use planning and wildlife maintenance: guidelines for conserving wildlife in an urban landscape. *Journal American Planning Association* 57 (3):313-323.

Case 5

Environmental Hazards on the US-Mexico Border: The Case of Ambos Nogales

Questions Explored:

*What are some of the most important geographical, socio-economic and historical patterns that characterize **bi-national urban regions** (twin cities) along the US-Mexico border?*

*Do the **ecoregions**, **watersheds** and **airsheds** of the region match the federal and state political boundaries, as well as other spatial patterns of topography, landscape and human habitation?*

*How does the **geographic nexus** involving topography, relief, human settlement, and gas or liquid contaminant dispersion patterns create a potentially hazardous situation for humans living in Ambos Nogales?*

*How do **human factors** exacerbate or mitigate vulnerability to natural or human-induced hazards such as risk of exposure to a chemical hazard?*

Key Learning Outcomes:

The learner will be able to:

Locate the Ambos Nogales region of the US-Mexico border and identify its major topographic, land cover, land use, and political subdivisions.

Identify some of the common land use, socioeconomic, historical and spatial characteristics of US-Mexico borderland communities.

Develop an understanding of the impacts caused by the differences between the spatial extent of natural physical processes and human political jurisdictions.

Discuss, using maps and imagery, the physical and human processes underlying the patterns of human vulnerability to exposure to a chemical hazard in the Ambos Nogales region.

Discuss policy and technical alternatives that could eliminate or mitigate the potential for environmental contamination to the population.

OVERVIEW

Location

This learning activity will examine human-environment interactions of **bi-national urban regions** (twin cities) along the US-Mexico border. The case looks specifically at Nogales, Arizona and Nogales, Sonora, together known as **Ambos Nogales**. The focus is on understanding the **vulnerability** of the inhabitants of these cities to a

---

Author:

Mark V. Finco and George F. Hepner
University of Utah

chemical emergency that might occur on the Mexican side and its potential impact on the US side. This case is particularly useful for assessing the peculiar problems of **trans-boundary and trans-national** environmental contamination issues.

Primary Objectives

The primary objective of this case is to demonstrate the interaction of the **dispersion** of a contaminant gas with the local patterns of topography and human settlement. It also illustrates how a **GIS** (Geographical Information Systems) might be used to model and simulate chemical emergencies so that urban planners and emergency personnel can respond more effectively and hopefully prevent such events. This case will also illustrate how the concept of **vulnerability** is applied in the study of **natural** or **human-induced hazards** within the earth systems sciences.

BACKGROUND

What are some of the most important geographical, socio-economic and historical patterns that characterize bi-national urban regions (twin cities) along the US-Mexico borderlands?

Figure 1. The US-Mexico Border Region with a focus on Arizona and Sonora.

U.S. interaction across the borderlands with Mexico has a long and tortuous history dating back to the colonial and later Independence era - remember the "Alamo" incident and the creation of the State of Texas and the Mexican-American War. During the American Civil War, the borderlands region of Mexico increased in importance as a major funnel for both smuggling and legitimate trade. Later, during the period leading up to the Mexican Revolution, cross-border incidents such as the infamous "chase for Pancho Villa" by the U.S. military added to the region's colorful history. Figure 1 shows the US-Mexico Border Region with a focus on Arizona and Sonora. Ambos Nogales is located on the US-Mexico border. It is the name used for both Nogales, Arizona in the US and Nogales, Sonora in Mexico.

Today, the **US-Mexico border** separates two very distinct nations at peace. Yet the border separates Mexico, a developing country with an average income of approximately $2,000-3,000 per year, and the United States of America, an economically developed nation with an average income of $18,000-20,000 per year.

Most twin cities along the US-Mexican border have a common population distribution pattern. Typically, t the Mexican city has a much higher population. Nogales, Arizona has a population of approximately 20,000, while more than 200,000 people live in Nogales, Sonora. It is not unusual for Mexican border cities to have a 10:1 differential in total population (see maps later in the case study).

In fact, the **borderlands** of Mexico in conjunction with the **Sunbelt** in the US have become one of the most politically and economically dynamic regions in the world. The sheer amount of interaction—movement of goods and people across it—surpasses any other in the world. Furthermore, most Americans do not realize that trade across the border with Mexico now makes it our third most important trading partner next to Canada and Japan. Unfortunately, this complex **bi-national relationship**, for better or worse, has given rise to a whole spectrum of perceptions about the borderlands and each other's country and people that often reflects quite disparate feelings about the ethnic, political, historical, socio-economic and demographic issues that bind us together. Some of the perceptions are true, but many are fallacious.

One of the most important industrial-economic developments in the region that is affecting all aspects of life has been the phenomenal growth of transnational manufacturing facilities called *maquiladoras*. Most of this growth has occurred since the 1965 passage of the **Mexican Border Industrialization Program**. Approximately 80 maquiladora plants involved in electronics assembly, plastics fabrication and automotive part production are located in Nogales, Sonora. Throughout the border region, there are more than 2,300 *maquiladoras* with over 600,000 employees.

The growth of *maquiladoras* in the border region has been stimulated by many factors including: lower labor costs, tariff and value-added tax differentials, and limited enforcement of environmental regulations. Geographic proximity of the transnational and local firms to the huge U.S. market has been a major attraction, as well as the various tax and investment advantages offered by the governments on both sides of the border (see South 1990; and Khosrow 1990). It is not surprising, therefore, that northern Mexico has the lowest unemployment rate in the country and attracts millions of migrants from poorer regions to the south.

Many migrants stay permanently or use the borderlands as a staging area to attempt eventual migration (legal and illegal) into the equally dynamically changing and growing **Sunbelt** of the United States (see Massey, Douglas S. 1987). Long term permanent residents of northern Mexico, or *Norteños* as they are called, share many cultural and linguistic patterns that add to the region's uniqueness. The language, for instance, includes many "Anglicisms" which some call *Spanglish* (and vice-versa). Long term interaction between the "Anglos" and "non-Anglos" in the region has created what Louis Casagrande calls one of the *"five nations of Mexico"*—**Mex-America**. You may want to read more about this unique region and its complex ethnic and settlement history by consulting Casagrande, Louis B. 1987—see also: Arreloa, D., Curtis, J. 1993 and Martinez, Oscar J. 1994.

Do the ecoregions, watersheds and airsheds of the region match the
federal and state political boundaries, as well as other spatial
patterns of topography, landscape and human habitation?

Natural regions of vegetation, soils, water drainage and air movement greatly affect the **human-spatial** characteristic of the region's landscape. **Topography** has been particularly influential in defining the urban development pattern in Nogales. Note how the urbanized area of Ambos Nogales extends along a narrow valley surrounded by a very steep and complex terrain. Primary transportation routes, such as the highway and the railroad, also follow the valley. The growth and expansion of residential and commercial land use also conform to the configurations of the landscape. The major centers of industrial land use are located in the southwest and extreme southern portions of the valley on the principal transportation routes that focus on the Nogales, Sonora border crossing.

Groundwater **aquifers** underlying the **Santa Cruz River** channel are the primary water sources for the citizens of both cities. The **watershed** and **airshed** boundaries are heavily influenced by the **landform** features and **relative relief** of the region. Finally, note how the national and state boundaries cut across the natural landscape in straight lines, often completely at odds with natural ecosystem and topographical boundaries.

Figure 2 is a map of the **ecoregions** of the border showing that Ambos Nogales is located in the **Sonoran Desert**. The Sonoran Desert is characterized by rugged, rocky terrain, with an average precipitation of only 15-30 cm per year. Typical vegetation includes cactus as the saguaro, ocotillo and prickly pear, the *Palo verde* tree and creosote bush.

Figure 2. Ecoregions along the Arizona/Sonora border.

THE HUMAN DIMENSION

The unique **topography, relative relief** and consequent **land use patterns** make exposure to the risk of contamination by liquid and airborne hazardous materials an issue of great concern to residents on both sides of the border. For one thing, the

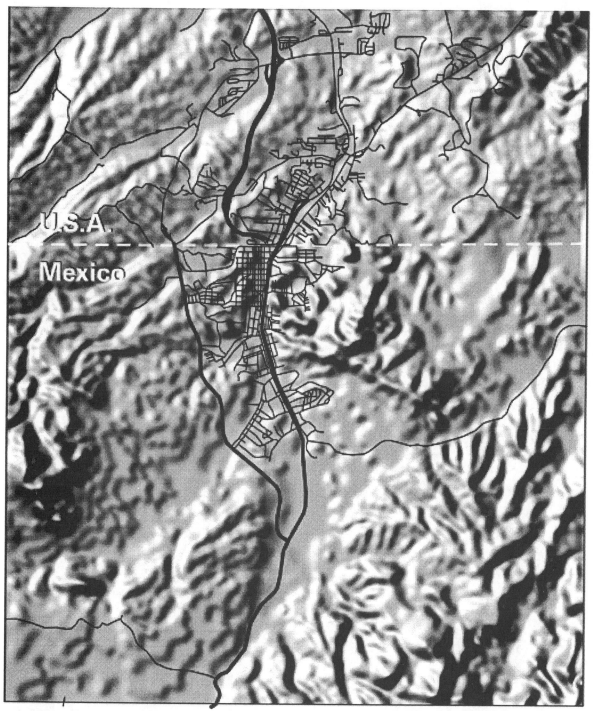

Figure 3. Above is a **digital elevation model** that shows the road network superimposed on a terrain model of the region.

approximate elevations of the major industrial zones in Nogales, Sonora are 1250-1300 meters while the elevations of the primary business and residential areas are somewhat lower at 1150-1200 meters. In the event of a toxic spill, **hazardous contaminants** released into the air would follow pathways from the higher elevation industrial to the lower elevation commercial and residential areas. Furthermore, the steep-sided valley configuration would tend to contain airborne contaminants within the populated area for a longer duration, thereby increasing the potential risk to human health. Also, the destination of any surface or sub-surface liquid contaminants would be the spatially constrained aquifers along the Santa Cruz River from which thousands of people farther north in Arizona pump their potable water.

Figure 4 is a map of the population density showing that the industrial sites are very close to and upslope from the areas of high population density. The population density is much higher in Nogales, Sonora than in Nogales, Arizona.

As noted earlier, most of the *maquiladoras* in Nogales, Sonora are located in the industrial zones at the southern end of the urbanized area (Figure 5). These industrial sites are potential sources of liquid and dense gaseous hazardous materials such as **LPG, carbon tetrachoride**, and **freon**. If human control and regulatory mechanisms should fail, for whatever reason, the probabilities of a severe **chemical emergency** are high. In sum, the **geographic nexus** of human settlement pattern and industrial activities in association with an unusual set of environmental conditions forms the basis for high **vulnerability** for specific groups of people.

It is interesting to note that unlike the United States, poorer residential areas on the Mexico side of the border are located upslope on the hillsides and wealthier neighborhoods are downslope in the valley bottom. In case of a hazardous materials spill, the upslope (poorer residential) locations may be in a more advantageous position in Mexico. This pattern is often reversed on the U.S. side of the border. This partially reflects differing patterns of urbanization in Latin America as compared to United States.

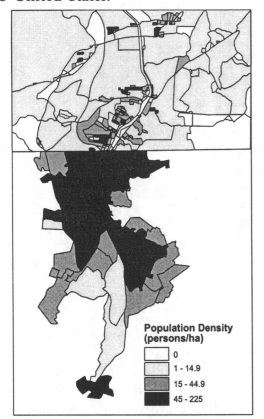

Figure 4. A population density map of Ambos Nogales.

54

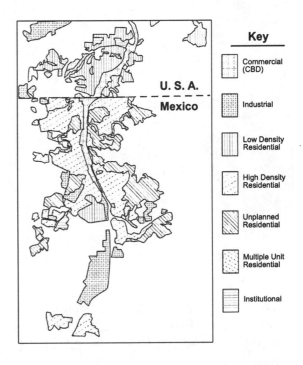

Figure 5. A map of land use patterns in the region with the sites of the industrial facilities shown.

How do human factors exacerbate or mitigate vulnerability to natural or human-induced hazards such as risk of exposure to a chemical emergency?

Although not the only factor, **gravity** is the major force acting upon liquid and gaseous materials released from industrial sites during an emergency. As noted earlier, the release of a liquid or gaseous contaminant from one of the industrial complexes in Nogales, Mexico would disperse toward the areas of highest residential density and commercial activity. The rate of dispersal and potential severity of a chemical emergency would depend on such factors as the amount and toxicity of the material released, its rate of dispersal and physical/chemical state (gaseous, particulate, or fluid), the potential for synergistic interactions with other events (e.g. earthquake, flood, fire), and local microclimatic and hydrological conditions at the time of dispersal.

Though the technical disciplines of environmental toxicology, geology and industrial hygiene have given us many sophisticated scientific tools and concepts for better understanding and analyzing toxic spills associated with modern industrial activity, the **actual capacity to respond** effectively and rapidly in a given chemical emergency is more often a function of **human-cultural management** issues and not just technical scientific knowledge or capacity. Consequently, the actual **vulnerability** of specific industrial workers or local residents within a given dispersion area such as the Ambos Nogales valley is often a function of less objective human factors such as:

- availability, training, readiness and effectiveness of local or off-site emergency response personnel;
- transportation and communications (time-distance relationships) for either "escape" by the victims or "response and mitigation" by appropriate authorities;
- local and regional (bi-national) politics; patterns of local land use planning, zoning and environmental regulation;
- social class and cultural perceptions of hazards;
- population density, health, age and socio-economic characteristics of residents in the areas concerned; and
- quality and maintenance of urban infrastructure systems including water, gas, transportation and electricity.

These varied factors express themselves on a given landscape in complex spatial patterns that are often quite difficult to integrate into a coherent whole. Mapping or modeling "potential emergency scenarios" with a GIS has been shown to be a particularly useful tool for planning for, preventing and mitigating such emergencies.

Figure 6 was produced from a computerized **geographic information system (GIS)** to integrate the socio-economic factors that contribute to the **vulnerability** of a community. Maps like these, that show vulnerability, can be integrated with **chemical dispersion models** to evaluate which groups of people might be most vulnerable to an accidental release of a dense gaseous contaminant. It also helps us assess the socio-economic relationships associated with those groups most likely to be affected—factors such as housing density and quality, poverty and social class, transportation access, availability of public services, access to health care, education, and even perceptions of risk.

In conclusion, effective response as well as analysis of potential environmental hazard scenarios requires integrating knowledge from many fields of study that span from the humanities to the social, physical and natural sciences. The **Earth systems science** (ESS) approach discussed in this book has attempted to show the advantages of an "integrative" systems approach to solving "real world" problems such as chemical contamination hazard risk. This system is particularly helpful in improving our understanding of the spatial relationships involved in such issues.

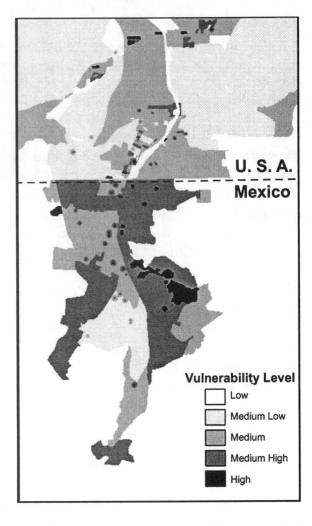

Figure 6. Composite spatial distribution of human vulnerability.

CONCEPTUAL QUESTIONS

Why are Ambos Nogales and other borderland Mexican cities such favorable geographic locations for large urban regions and industrial complexes such as *maquiladoras*?

What do you think contributes to the great variation in population totals in these "sister cities"? Remember that the differences are often more than 10:1 from one side of the border to the other.

What physical and cultural characteristics are unique to the Nogales region, in contrast to those it shares with other U.S.-Mexico border cities? Is there such a thing as a "borderland" urbanization pattern?

How would a hazardous contamination "event" be different if this area had flat topography, more precipitation and was not located on an international border? Would the hazard be greater or less? Why?

What political, economic, historical, ethnic, or socio-cultural patterns on either side of the border exacerbate or mitigate the potential for such events?

What do you think policy makers on either side of the border can do to reduce the chances of such accidents? How can they best prepare to deal with such emergencies? Think about possible solutions in terms of who is "least" and "most" vulnerable?

SOME RECOMMENDED READINGS

Books:

Arreloa, D., Curtis, J. 1993. *The Mexican Border Cities: Landscape Anatomy and Place Personality.* Tucson: University of Arizona Press.

Barry, T. 1994. *The Challenge of Cross-Border Environmentalism.* Albuquerque: Resource Center Press.

de Blij, Harm J. 1997. *Geography: Realms, Regions, and Concepts.* John Wiley & Sons, Inc.

Heyman, Josiah McConnell. 1991. *Life and Labor on the Border: Working People of Northeastern Sonora, Mexico 1886-1986.* Tucson: University of Arizona Press.

Khosrow, Fatemi. 1990. *The Maquiladora Industry: Economic Solution or Problem?* New York: Praeger.

Martinez, Oscar J. 1994. *Border People: Life and Society in the US-Mexico Borderlands.* Tucson: University of Arizona Press.

Murck, Barbara W., Brian J. Skinner, and Stephen C. Porter. 1996. *Environmental Geology.* New York: John Wiley & Sons, Inc.

Murck, Barbara W., Brian J. Skinner, and Stephen C. Porter. 1997. *The Dangerous Earth: An Introduction to Geologic Hazards.* New York: John Wiley & Sons, Inc.

Skinner, Brian J. and Stephen C. Porter. 1995. *The Blue Planet: an Introduction to Earth System Science.* New York: John Wiley & Sons, Inc.

Skinner, Brian J. and Stephen C. Porter. 1995. *The Dynamic Earth: an Introduction to Physical Geology* (3rd edition). New York: John Wiley & Sons, Inc.

Smith, Keith. 1996. *Environmental Hazards: Assessing Risk and Reducing Disaster (2nd Edition).* London: Routledge.

Stoddard, Ellwyn R. 1987. *Maquila: Assembly Plants in Northern Mexico.* University of Texas El Paso, Texas Western Press.

Strahler, Alan and Arthur Strahler. 1996. *Introducing Physical Geography: Environmental Update.* New York: John Wiley & Sons, Inc.

Texas Center for Policy Studies. 1990. *Overview of Environmental Issues Associated with Maquiladora Development along the Texas-Mexico Border.* Austin, Texas.

US Environmental Protection Agency. 1996. *US/Mexico Border XXI Program: Framework Document.* EPA160-D-96-001. Washington, D.C.

West, Robert C. and John P. Augelli. 1989. *Middle America: Its Lands and Peoples.* Englewood Cliffs, New Jersey: Prentice Hall.

Articles:

Angotti, Thomas. 1987. Urbanization in Latin America: towards a theoretical synthesis. *Latin American Perspectives*, 14(2): 134-56.

Balcazar, H; Denman, C; Lara, F. 1995. Factors Associated with Work-related Accidents and Sickness among Maquiladora Workers - the Case of Nogales, Sonora, Mexico. *International Journal of Health Services.* 25 (3): 489-502.

Bowen, M., T. Kontuly and G. Hepner.1995. Estimating Maquiladora Hazardous Waste Generation on the US/Mexico Border. *Environmental Management.* 19(2): 281-296.

Casagrande, Louis B. 1987. The Five Nations of Mexico. *Focus* (American Geographical Society). Spring: 2-9.

Davis, A. R. Altamirano. 1989. Hazardous Waste Management at the Mexican-US Border. *Environmental Science Technology.* 23 (10): 1208-1210.

Ford, Robert E. and Valerie M. Hudson. 1992. The USA and Latin America at the end of the Colombian Age: How America 'Cut the Atlantic Apron Strings' in 1992. *Third World Quarterly.* 13(3): 441-462.

Hepner, G., and M. Finco. 1995. Modeling Dense Gaseous Contaminant Pathways Over Complex Terrain Using a Geographic Information System. *Journal of Hazardous Materials.* 42: 187-199.

Massey, Douglas S. 1987. Understanding Mexican migration to the United States. *American Journal of Sociology.* 92(6): 1372-1403.

Sargent, Charles S. 1989. The Latin American city. In *Latin America and the Caribbean: A Systematic and Regional Survey.* (Second Edition). By Brian W. Blouet and Olwyn M. Blouet. John Wiley & Sons. Pages 172-216.

South, Robert E. 1990. Transnational 'Maquiladora' location. *Annals of the Association of American Geographers* 80(4): 549-570.

Case 6 Saline Lakes and Global Climate Change

Questions Explored:

What are some of the world's most well-known saline lakes, how were they formed, and what can they tell us about the Earth's geologic and climatic history?

Why are saline lakes important?

How can the saga of the Aral Sea serve as an example of the effects of human intervention with sensitive natural systems?

Key Learning Outcomes:

The learner will be able to:

Locate at least eight saline lakes on a world map and list their common attributes.

Describe at least three different ways in which saline lakes can be formed and give an example of each.

Explain how saline lakes can serve as a monitoring system of global climate change.

List at least four reasons why saline lakes are valuable natural and/or economic resources.

OVERVIEW

Location

Saline lakes are found in dry climate zones of every continent, and cover almost as much of the earth's surface as do freshwater lakes. The volume of water contained in the Earth's saline lakes is about 104,000 cubic kilometers, compared to a volume of 125,000 cubic kilometers for freshwater lakes (Vallentyne 1972). Many salt lakes in extremely arid regions are **ephemeral**, containing water only during part of the year or only during years of exceptionally high precipitation. A good example of this is Lake Eyre in Australia. During the rare occasions when Lake Eyre is at its maximum depth of about six meters, it is the seventh largest salt lake in the world. Much of the time, however, evaporation exceeds precipitation so much that the lake bed is completely dry. The majority of permanent saline lakes are found in semi-arid and subhumid climate zones where evaporation is lower.

Primary Objectives

The overall goal of this case study is to develop an appreciation of an important, often-overlooked ecosystem - the saline lake. This goal can be divided into three primary objectives:

Author:

Sandra A. Zicus
Salt Lake City, Utah

1.   to illustrate how these sensitive ecosystems can serve as windows to the Earth's geologic and climatic past,
2.   to increase awareness of the natural and economic importance of saline lake ecosystems, and
3.   to demonstrate how human intervention with a saline lake ecosystem - the Aral Sea - has led to an ecological disaster.

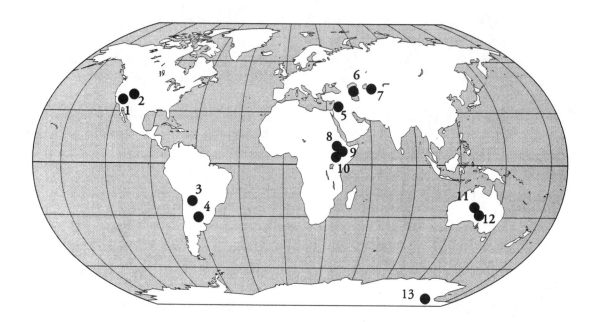

Figure 1. World map with the following major lakes located - 1-Mono Lake (California, USA), 2-Great Salt Lake (Utah, USA), 3-Laguna Colorada (SW Bolivia), 4-Laguna Mar Chiquita (Cordoba, Argentina), 5-Dead Sea (between Jordan & Israel), 6-Caspian Sea (boundary S.E. Europe and S.W. Asia), 7-Aral Sea (border between Kazakhstan & Uzbekistan,), 8-Lakes Natron, 9-Magadi, and 10-Elementeita (East Rift Valley - East Africa), 11-Lakes Amadeus and 12-Eyre (central Australia), and 13-Lake Vanda (Wright Valley, Antarctica).

BACKGROUND

What is a saline lake, and what attributes do saline lakes share?

To be considered a true saline lake, a lake must contain a concentration of dissolved minerals greater than 3% by weight. For comparison, the oceans average about 3.5%. Saline lakes around the globe tend to share certain similarities:

1.   They occur in **arid, semi-arid, or subhumid environments** where average evaporation equals or exceeds precipitation, and are usually located in the interior of a continent or in the rain shadow of a mountain range.
2.   They are mostly **terminal** (having no outlet), and have a high dissolved

salt content (more than 3% by weight). The salt concentration may be the result of evaporation of a much larger body of fresh water or of the closing off of an arm of an ocean due to tectonic forces or sea level change.

3.   All have **highly specialized ecosystems** that are sensitive to changes in water level and salinity.

4.   All are **sensitive to regional changes**, whether natural or human-induced, **in climate or water balance.** These changes lead to water level fluctuations with corresponding effects on the ecosystems, and on human populations and their constructions.

What are some of the world's most well-known saline lakes, how were they formed, and what can they tell us about the Earth's geologic and climatic history?

The **Caspian Sea**, the world's largest inland body of water, covering some 143,000 square miles, is located on the boundary between southeastern Europe and southwestern Asia, and is surrounded by the countries of Azerbaijan, Iran, Kazakhstan, Russia, and Turkmenistan. The surface elevation of the lake averages 94 feet below sea level, and it has an average maximum depth of 3,360 feet. Unlike most salt lakes, which begin as freshwater lakes and gradually become saltier, the Caspian began as a remnant of an ancient ocean, the Tethys Sea. During the mid-Miocene, an arm of the Tethys, known as the Paratethys, was isolated by mountain building as Africa collided with Eurasia. This formed the basin that now contains the Caspian, the Black, and the Aral Seas.

Because of the large input of fresh water from the Volga and five other rivers (Ural, Emba, Gurgan, Atrek, and Kura), the Caspian has become less salty over time. Now, only certain portions are salty enough to be considered saline. Currently, the overall salinity of the sea is 1.3% and dropping. Since 1978, the lake level has risen rapidly (about 8 feet) due to increased regional precipitation and decreased evaporation. This has led to serious flooding, which has drawn international attention. Like the Great Salt Lake in northeastern Utah, the Caspian is located in a flat basin and has large cities along its coastline. Flooding has damaged buildings and roads, and is threatening offshore oil rigs. International studies are ongoing to determine the causes of the rapid level changes, emergency dikes are being constructed, and the town of Astara in Iran has put three pumps into operation in an attempt to carry floodwaters out of town.

Perhaps the most famous salt lake in the world is the **Dead Sea**, on the border between Jordan and Israel. Located in the lowest point of the Great Rift system, the lake has an average surface level of about 1300 feet below sea level, making it the lowest surface point on Earth. The climate around the Dead Sea is extremely hot and arid. Maximum summer temperatures average slightly over 100° F, and the region receives less than 4 inches of rain per year. The basin receives most of its water from the Jordan River, which enters from the north. The lake is divided into two sections by a low ridge known as the Lisan Straites. The deepest part of the lake (about 1300 feet) is in the northern basin, while the smaller southern basin is very shallow (less than 20 feet deep). Between 1964 and 1976, the southern basin was dammed and converted into evaporation ponds for the production of salts.

The Dead Sea has a salinity ranging from 28 to 29%. The chemical composition of the water has allowed for the development of economically important mineral industries producing products such as potash, bromine, gypsum, common salt, and natural beauty and health products. The area is also renowned as a health resort.

Currently, the Dead Sea is about 47 miles long and 10 miles wide, with an average surface area of about 386 square miles. Beginning in the late 1930s, numerous water development projects were started which greatly decreased the amount of water reaching the lake from the Jordan River. Between the late 1940s and the present, the amount of river water reaching the lake has decreased by three-quarters, leading to decreased surface area and greatly increased salinity. Another effect of the declining lake level has been an increase in the flow of groundwater from surrounding areas toward the lake. In an effort to address the environmental and economic impacts of the lake level decline, a proposal has been made to construct a canal linking the Mediterranean to the Dead Sea.

In North America, the **Great Salt Lake** in Utah and **Mono Lake** in California are two of the most well-known saline lakes. The story of the Great Salt Lake goes back at least 18 million years, to the geologic event that created the basin it occupies. The continental crust underlying much of the western United States has been slowly stretching in an east-west direction, and has formed a region that is known as the Basin and Range Province.

This stretching has caused the brittle rocks of the Earth's crust to fracture along numerous north-south faults, dividing the crust into elongate blocks. Between the faults, the blocks have sunk and tilted as the crust continued to stretch, resulting in alternating parallel valleys (basins) and mountains (ranges). The valleys created by the stretching have been partially filled by sediments eroded from the mountains, forming flat-floored basins. The Great Salt Lake lies within three of these parallel, interconnected valleys, with the intervening uplifted ridges of rocks forming islands and the promontories that extend into the lake. The lake occupies the lowest part of the Great Salt Lake Valley.

Another event critical to the formation of the Great Salt Lake was a series of volcanic eruptions that took place in what is now southern Idaho. Lava flows from these eruptions began to divert the course of the Bear River southward into western Utah through the Cache Valley about 50,000 years ago. This diversion was complete by 30,000 years ago, and a lot more water was entering the basin.

At this time, the climate was much cooler and wetter than it is today. There was abundant rainfall and lower rates of evaporation. Increased snowfall, combined with the cooler weather, resulted in the formation of glaciers on the higher peaks of the Wasatch Range. These glaciers, with their seasonal melting, supplied additional water to the valley. Numerous freshwater lakes which had formed in depressions in the basin gradually deepened. Around 25,000 years ago, they merged to form giant Lake Bonneville. In its heyday, Lake Bonneville was 1,200 feet deep and covered about 20,000 square miles.

Around 14,500 years ago, the waters of Lake Bonneville broke through a natural dam to the north at Red Rock Pass in Idaho. A catastrophic flood roared down the Snake River Canyon, and the lake level dropped by 350 feet in less than a year. The lake remained at this new level, known as the Provo level, for at least 500 years.

At the end of the Pleistocene, the climate became increasingly warm and arid. By 12,000 years ago, the lake level had fallen another 1,000 feet. As water continued to evaporate from the lake, the salts were concentrated and, by approximately 10,000 years ago, the lake became saline. Currently, the Great Salt Lake covers an average area of 1500 square miles and has a salinity that ranges between 5% and 27%, depending on the season, long-term weather patterns, and the location within the lake.

Mono Lake, like the Great Salt Lake, was a much larger freshwater lake during the Pleistocene. It is located at the boundary of two different geographic provinces - the Great Basin and the Sierra Nevada mountain range. Five major streams bring fresh water into the lake.

In 1941, an aqueduct was completed and the City of Los Angeles started to divert water from the basin. By 1991, the lake level had dropped 43 vertical feet, the lake volume was halved, and the salinity doubled to about 9%. Mono Lake currently has a surface area of about 66 square miles. Seventeen thousand acres of former lakebed, encrusted with alkali salts, are now exposed to wind erosion. Strong winds blowing through the valley carry dust laden with sulfates, selenium, arsenic, and other toxic substances up to 100 miles.

Mono Lake has an ecosystem similar to that of the Great Salt Lake, based on algae, brine shrimp, brine flies, and a wide variety of birds such as California gulls, eared grebes, and Wilson's phalaropes. Mono, along with the Great Salt Lake, has been designated a Western Hemisphere Shorebird Reserve, and is twinned with Laguna Mar Chiquita in central Argentina.

Lowering water levels and increasing salinity are a major threat to the sensitive ecosystem. In early 1995, a court order required Los Angeles to stop withdrawing water from the streams that feed Mono Lake until the lake level rises at least 16 feet.

Laguna Mar Chiquita, in the province of Cordoba, Argentina, is a large, shallow, saline lake with a surface area of around 2,300 square miles, and surrounded by another 1,500 square miles of wetlands. Three major rivers - the Suquía, Xanaes, and Dulce - feed the lake, and 25% of Argentina's bird species can be found at the lake or in the adjacent wetlands and plains. More than 70,000 Chilean flamingos nest at the lake, and 1,000 Andean flamingos travel down from the frigid Altiplano for the winter. Because of the lake's importance as a wintering ground for about 500,000 Wilson's phalaropes, the lake was paired with Mono Lake and the Great Salt Lake through the

Western Hemisphere Shorebird Reserve.

In 1977, because of prolonged wet weather, the lake level started rising. By the mid 1980s, the city of Miramar, with a population of 3,000 people, had to be relocated farther from the lake shore. The water salinity has dropped to about 1/10 of its normal level, causing a major decrease in the brine shrimp population and flooding many of the nesting islands and mudflats that are of importance to gulls and shorebirds.

In Antarctica is **Lake Vanda**, which is located in the Wright Valley, one of the so-called Dry Valleys near the Ross Sea. These large, glacially-formed valleys extend from the Transantarctic Mountains at an elevation of almost 10,000 feet to the coast. They are called the Dry Valleys because they have had no rain or snow for millions of years, and the valley floors are not permanently covered with snow and ice.

Lake Vanda is formed from summer snowmelt carried from the Polar Plateau by the Onyx River. A small lake, measuring only 5 miles long by slightly more than a mile wide, with a maximum depth of more than 250 feet, it is permanently covered by about 13 feet of ice. Because of solar heating through the ice, the water temperature at the lake bottom averages around 78° F with some areas having temperatures as high as 113° F! The lake is also divided by salinity and density - the surface is freshwater, while the salinity at the lake bottom is about 6%. The density difference keeps the lake waters from mixing.

In recent years, the level of Lake Vanda has been rising by about 3 feet a year. The volume of meltwater carried by the Onyx River has more than doubled in the last ten years. There is speculation that global warming may be causing increased melting of the Antarctic glaciers.

In East Africa's Great Rift Valley are numerous salt lakes that are quite different. These African lakes are called **soda lakes** because, instead of sodium chloride, they contain mainly sodium carbonate, supplied by hot springs and by erosion of the surrounding carbonate-rich terrain. Due to their unique chemistry and biota, the lakes sometimes appear bright red from the air. **Lakes Natron, Nakuru, Magadi**, and **Elmenteita**, along with several other soda lakes in the region, support approximately one-half the world's flamingo population. The birds feed on algae and zooplankton from the lakes, and nest on dry portions of the lake bed where the harsh climate helps keep them safe from predators.

Lake Nakuru in Kenya is famous for hosting up to two million lesser flamingos, and is a major tourist attraction. Since 1993, however, the lake has been rapidly shrinking due to a prolonged drought, and the flamingos are moving to other lakes in the region. The drought may be part of a natural cycle, but there is concern that it could be related to a rapidly expanding population, intensified cultivation, and increased deforestation in the surrounding highlands.

The **Aral Sea**, on the border of Kazakhstan and Uzbekistan, provides a classic example of human-caused ecological disaster. Once the fourth largest lake in the world, the slightly-brackish Aral provided a moderating influence on the climate of the region, served as a major transportation corridor, and supported a thriving fishing industry that employed more than 70,000 people. The Aral Sea and its current state will be discussed in more detail in the section "The Human Dimension."

The present lake was formed during the late Pleistocene when the Amu Dar'ya River, which was flowing west toward the Caspian Sea, changed course to flow north into the Aral basin, bringing enough water to form a large lake. The basin, however, has existed for at least 2 million years, and has repeatedly flooded and dried up, depending on the climate.

Why are saline lakes important?

As should be clear from the above descriptions, salt lake ecosystems are extremely valuable in both ecological and economic terms. Many of them serve as important nesting and/or feeding and resting grounds for various species of waterfowl. The African soda lakes and the lakes of the South American altiplano are essential sites for flamingos. More than 500,000 Wilson's phalaropes migrate annually between the salt lakes of the western United States and Argentina and Bolivia. The Great Salt Lake in Utah is such an important site for birds that it has been designated as a Western Hemisphere Shorebird Reserve. Millions of shorebirds of more than 30 species use the Great Salt Lake ecosystem every year.

Saline lakes also provide a variety of economic benefits, some direct and some indirect. Direct economic benefits include valuable mineral resources such as sodium chloride used as table salt, road salt, and in water softeners; potassium sulfate for fertilizers; magnesium chloride and magnesium metal; and sodium sulfate which is used in the production of ceramics, paper and detergents. In lakes where the water is not too saline, commercial fishing is often an important economic resource. Brine shrimp eggs (Artemia spp.) are also harvested from salt lakes in many countries. The brine shrimp industry in Utah earns as much as 30 million dollars per year, and the eggs are used for aquaculture around the world and as food for aquarium fish.

As indirect benefits, the larger lakes provide a moderating influence on regional climate, decreasing temperature extremes and increasing relative humidity. This in turn lengthens the growing season of the region. Since salt lakes generally occur in dry climates, increased precipitation is another important benefit of the presence of the lake. Many of the lakes are also valuable for use as recreation spots or health resorts.

THE HUMAN DIMENSION

How can the saga of the Aral Sea serve as an example of the effects of human intervention with sensitive natural ecosystems?

In 1961, the Aral Sea had a surface area of 25,830 square miles and was slightly brackish, with a salinity of about 1%. By 1995, because of water diversion from the two principal rivers, the Amu Dar'ya and the Syr Dar'ya, for irrigation of cotton and rice fields, the Aral had lost 75% of its volume and 50% of its surface area, the shoreline had retreated up to 75 miles, and the salinity had risen to 3%.

As a result of the higher salinity and the retreating shoreline, the fishing industry is gone, leaving more than 70,000 people unemployed. The climate has gotten drier, and the daily and seasonal temperature ranges more extreme. Summers average about 2° C warmer and winters 2° C colder. Average relative humidity has decreased, and there has been an increase in the number of arid days. Dust storms have become so large and extreme that they have been observed from space. A large dust storm can transport more than 1.5 million tons of dust and salts, including heavy metals and other toxins, for distances of up to 400 km. The salt level in atmospheric precipitation increased six-fold between 1969 and 1979. Both of these factors have led to increased soil salinization and decreased crop productivity. As soil salinity increases, more irrigation water is needed to continue to grow crops. More water must be diverted from freshwater rivers and streams. As this irrigation water evaporates on the fields, salts from the water are added to those already in the soil, groundwater, and rivers, increasing the salinity of all and creating a positive feedback loop.

In addition, because of later spring frosts and earlier autumn ones, the growing season in the region has been shortened by about a month. The growing season around the Aral is now too short for successful cultivation of cotton.

Lowering of the water table in the deltas surrounding the lake has led to **desertification** and loss of wetlands and riparian habitat, with a corresponding increase in **xerophytes** and **halophytes**, plants that can survive in dry or relatively salty conditions. More than 75% of the 173 species of animals dependent on the lake are locally extinct, and environmentally-induced human health problems have increased dramatically.

These human health problems result from a combination of factors including wind transport of pollutants exposed on the dry lake bed and the contamination of ground water by salts and industrial and agricultural pollutants. According to a 1995 *Health & the Planet* report (Vol. 4, No. 2), regional health studies conducted by Dr. Oral Ataniyazova of Karalkapakstan, a semi-independent republic of Uzbekistan, indicate that, "During the past 10-15 years, kidney and liver diseases, especially cancers, have increased 30-40 fold, arthritic diseases by 60-fold and chronic bronchitis by 30-fold." In Muynak, a former port city at the southern end of the Aral Sea, the annual death rate is reported to be around 100 per 1,000 people. Over 90% of the 700,000 women living in Karakalpakstan are anemic, a condition caused, according to local doctors, by drinking water that is polluted by salts and concentrated chemicals from run-off from the cotton fields.

The water is still being diverted, and the lake is still shrinking. In 1993, an international agreement aimed at improving conditions in the region was signed by all five of the watershed nations surrounding the Aral Sea. By 1997, however, the plan had not accomplished much. There are still conflicting demands for water use among the upstream countries of Kyrgistan and Tajikstan, which want to use the water of the Amu Dar'ya and the Syr Dar'ya to produce energy, and the downstream countries of Kazakhstan, Turkmenistan, and Uzbekistan, which need the water for continued irrigation if the agricultural base of the region is to survive.

Various strategies have been proposed to restore water to the Aral, such as building pipelines to bring in water from other rivers, improving irrigation efficiency to decrease the amount of water needed, reducing economic dependence on cotton by switching to manufacture of artificial fibers, thus reducing the need for irrigation, decreasing the area planted in rice, and intensifying reforestation efforts in the region. Thus far, however, no multi-national agreement on the best way to proceed has been reached.

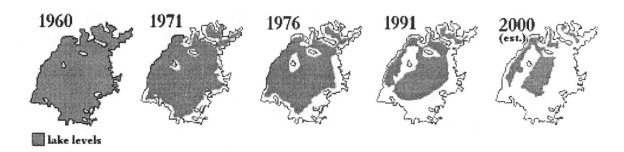

Figure 2. The above diagram is a chronology of Aral Sea changes from 1960 through the year 2000 (estimated). Since 1960 the Aral Sea has lost two thirds of its volume and half of its surface area (modified from data obtained from the United Nations Environmental Programme).

CONCEPTUAL QUESTIONS

Since most saline lakes occur in closed basins (having no outlet), how do prolonged periods of exceptionally wet or exceptionally dry weather affect the lakes? What about significant long-term changes in mean temperatures?

What are the principal factors that affect saline lake ecosystems when lake levels change significantly? Why?

What similarities are found in the geographic settings of the Caspian Sea, the Great Salt Lake, and Laguna Mar Chiquita? How does the basin geography affect human settlement patterns near the lakes? What factors need to be considered when developing a management plan for lake use and for land use near the lake shores?

What can we learn about the Earth's past climate history by studying saline lakes? How can this information help us predict the impacts of future global or regional climate change?

SOME RECOMMENDED READINGS

Books:

Greer, Dean C., ed. 1977. *Desertic Terminal Lakes.* Proceedings for the International Conference on Desertic Terminal Lakes.

Gwynn, J. Wallace, ed. 1980. *Great Salt Lake, a Scientific, Historical and Economic Overview.* Utah Geological and Mineral Survey Bulletin 116, Salt Lake City, Utah.

Hammer, U.T. 1986. *Saline Lake Ecosystems of the World.* University of Saskatchewan, Saskatoon, Canada.

Zicus, Sandra, ed. 1997. *The Great Salt Lake Story: An Interdisciplinary Activity Guide.* Utah Museum of Natural History, Salt Lake City, Utah.

Articles:

Hinrichsen, Don. 1995. Requiem for a Dying Sea. *People & the Planet Magazine,* 4, No. 2.

Kairu, J.K. 1994. Pesticide Residues in Birds at Lake Nakuru, Kenya. *International Journal of Salt Lake Research,* p. 31, 3, No. 1.

Kotlyakov, V.M. et al. 1992. Concept for Preserving and Restoring the Aral Sea and Normalizing the Ecological, Public Health, and Socioeconomic Situation in the Aral Region. *Post-Soviet Geography,* pp.283-295, 33, No. 5.

Micklin, Philip P. 1992. The Aral Crisis: Introduction to the Special Issue. *Post-Soviet Geography,* pp.269-282, 33, No. 5.

Murimi, S.K. 1994. Falling water-levels in Saline Lakes of the Central Rift Valley of Kenya: The Case of Lake Elmenteita. *International Journal of Salt Lake Research,* p. 65, 3, No. 1.

Murzayeve, E.M. 1992. Research on the Aral Sea and Aral Region. *Post-Soviet Geography,* pp. 296-314, 33, No. 5.

Williams, W.D. and Sherwood, J.E. 1994. Definition and Measurement of Salinity in Salt Lakes. *International Journal of Salt Lake Research,* p. 53, 3, No. 1.

Zekster, I.S. 1995/96. Groundwater Discharge into Lakes: A Review of Recent Studies with Particular Regard to Large Saline Lakes in Central Asia. *International Journal of Salt Lake Research,* pp. 233-249, 4, No. 3.

Case 7
Landscape and Life along the East African Rift: the Virunga Mountains, Rwanda

Questions Explored:

Where is the East African rift system and how was it formed?

What types of landforms, landscapes and drainage patterns can be observed in the East African rift region that are related to tectonism and volcanism?

How are vegetation, climate and soils influenced by the underlying geomorphology, geology, and hydrology?

How have humans interacted with this landscape, and what are the implications for sustainable development?

How are human vulnerability and risk—such as land degradation—monitored and mitigated using "earth systems science" tools and techniques such as GIS / RS and GPS?

Key Learning Outcomes:

The learner will be able to:

*Locate the **Eastern African rift system** on a map, and identify its major subdivisions and key features; e.g., lakes, rivers, faults, rifts, and volcanoes in relation to the broader **Afro-Arabian rift system**.*

*Identify from maps, photos and remotely sensed imagery some of the **tectonic, volcanic, and hydrological features** typical of the Western Rift Valley in the vicinity of the **Virunga Mountains** of Rwanda, Uganda and Zaire.*

*Define, discuss, and identify on maps, photos and remotely sensed imagery some of the **ecological/ biospheric patterns** typical of this landscape; e.g. altitudinal zonation of vegetation, micro-climate patterns, and soils.*

*Discuss and evaluate social policy and scientific findings regarding **sustainable development** issues, particularly **land degradation** within the region.*

OVERVIEW

Location
This learning activity will explore the intricate functional linkages that exist between "land and life" in the **Virunga Mountains** of Rwanda. The Virunga Mountains are a region where D.R. Congo, Rwanda, and Uganda meet. This area also encompases the broader region known as the **East African rift system** (EARS). The Virunga mountains belong to the **Western Rift Valley**—itself a small segment within the much larger continental-scale **Afro-Arabian rift system** which extends from Turkey to Mozambique. The region has a spectacularly beautiful landscape where one can observe the primal

---

Author:

Robert E. Ford
Westminster College of Salt Lake City

forces of Nature at work. Many people also know it as the last bastion of one of Earth's most endangered species—the **mountain gorilla**. This awe-inspiring primate has been made famous by the selfless work of Dian Fossey, George Schaller, and many others.

Primary Objectives

The primary objective of this case is to show how **tectonism** and **volcanism** have molded the **landscape** around the **Virunga Mountains**, and how those geological processes have influenced the distribution of plants, water, soil, animals and other natural resources. Humans have also interacted intensely with this landscape. Many facets of life reflect the underlying physical realities: land use and settlement patterns, agroecological systems, water resource management practices, population distribution and growth, health, economic development and even, to a degree, politics.

Most of the world today has heard about Rwanda because of the immense human tragedy and genocide that occurred in 1994, which spilled over into the adjoining countries bringing insecurity, political instability and environmental degradation. Rwanda is one of Africa's most densely settled rural areas. Some policy scientists and thinkers concerned with the issue of environmental security see in the Rwandan tragedy clues to what might possibly be the indicators of "optimal upper limits" to human population density on the land (given certain resource and sociocultural constraints).

The key question for students of the region is the following: have the limits to sustainable development been exceeded? If so, can we measure and predict these limits and hopefully prevent similar environmental or human-induced catastrophes? This case raises significant questions regarding the human dimensions of global change (HDGC) and its implications for sustainable development in the tropics as well as the whole of Planet Earth.

BACKGROUND

What is the East African rift system and how was it formed?

Many people still use the term **The Great Rift Valley** for the East African rift system after the name given it by one of Africa's great geologists, J.W. Gregory (Figure 1). The section of the rift that runs across Kenya and Tanzania is still known as the **Gregory Rift Valley** (GRV) in his honor. His book *The Great Rift Valley*, published first in 1896 and republished by Frank Cass (1968), is still a wonderful account by a fine observer of Africa's most spectacular landscape.

Today, earth scientists refer to the entire feature as a **rift system** because it is indeed a complex of tectonic and volcanic features that varies from one area to another. According to current understanding of plate tectonic theory, the EARS is a true **continental intra-plate rift,** as compared to the **mid-Atlantic rift** which is an **oceanic** one. Along the rift zone continental rupture is occurring and is particularly evident at the famous "triple junction" where the **Red Sea rift** (RSR) and **Ethiopian Rift Valley**

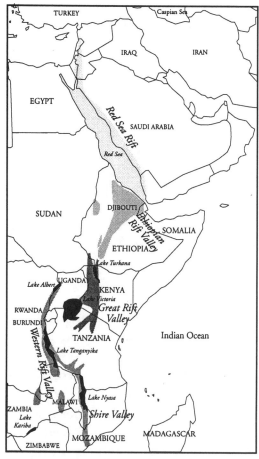

Figure 1. The Afro-Arabian rift system.

(ERV) intersect the **Gulf of Aden graben (GAG)** which is part of the **Mid-ocean Ridge and Central Graben** (Figure 1). Along the EARS new material from deep within the earth's mantle wells up from **hotspots** to bulge and thin the Earth's crust, producing both tensional and compression faulting and volcanism. Most of Africa's coast, particularly along the Red Sea is an excellent example of a **passive continental margin**.

Compared to the African continent as a whole, the East African rift system is a relatively new landscape. Africa was once part of the ancient **Gondwana** landmass that broke away from the supercontinent **Pangea**. Consequently, most of Africa is a very old **continental shield** that has been heavily eroded and planed down. The longterm erosion has produced a continental landscape with conspicuously large sediment-filled basins separated by low, rounded and heavily eroded uplands, e.g. the **Sudd** basin.

What types of landforms, landscapes and drainage patterns can be observed in the East African rift region that are related to tectonism and volcanism?

Some areas are remnants of ancient orogenic features called **cratons**. These tectonically stable zones contain very old rocks, among which are found some of the richest mineral areas of the world; e.g. in the **Transvaal** of South Africa. Rock outcrops of **Pre-Cambrian age** are quite common, as are sedimentary rocks of **Cretaceous** or **Carboniferous** age. In contrast, the **volcanic rocks** of East Africa are quite recent, dating from **Tertiary** and **Quaternary** periods. The **Atlas** and **Ruwenzori** mountains are two rare examples of zones where relatively new mountain-building or **orogeny** is occurring in Africa.

Tectonic-related Landforms and Surface Hydrology

At the local and regional-level, many types of **tectonic**-related **landforms** can be observed. Faulting, fracturing and subsidence of the Earth's crust have produced **step faults**, **horsts** and **grabens**. Look at the generalized cross-section across East Africa in **Figure 2** to see some of the effects of **uplift** and subsequent **faulting** on the region. Note that several of the "valleys" which many assume to be true grabens do not fit the classic definition—a down-thrown block of land between two parallel faults. Instead, some sections of the rift have produced **half-grabens**—cases where only one major fault is involved, resulting in an "asymmetrical" depression where the most prominent **escarpment** may be found on one side of the depression alone. **Figure 3** illustrates the half-grabens found in the Lake Tanganyika basin.

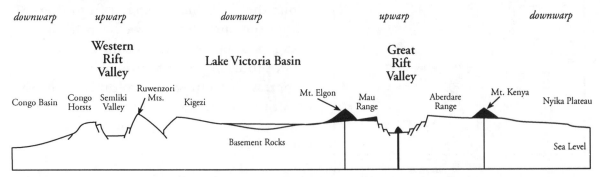

Figure 2. A cross-sectional diagram of East Africa from the Kenyan coast to Congo Basin showing zones of up and down warping and faulting as well as major landform features: rift valleys, volcanoes, escarpments and lakes.

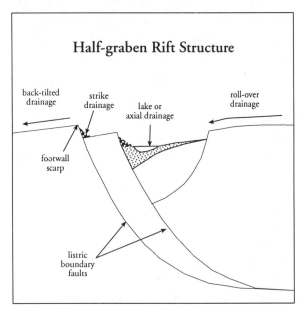

Figure 3. Diagram of half-grabens and drainage disruptions on Lake Tanganyika (modified from *Sommersfield, 1991*).

In some other cases, the classic parallel double-fault system is quite evident and recent (note the satellite images of **Lake Albert** and **Lake Edward in Virtual Tour I** which show the "straight-line" shores and parallel escarpments). **Lake Malawi**, on the other hand, has experienced complex multiple faulting and rifting over a longer time and exhibits no current volcanic activity.

One of the most dramatic impacts on the African landscape is the effect of tectonism and volcanism on surface hydrology. Lakes and rivers were **deranged** and wetlands and swamps were created through "drowning" of river valleys. As you can imagine, when rifting began it tilted, bulged, and depressed different areas causing some rivers to **capture** parts or all of adjacent **drainage systems**. Note the **Nyabarongo** river (**Figure 4**) which captured the drainage out of the **Mukungwa watershed**—which includes **Lakes Bulera** and **Ruhondo** and the **Rugezi** swamp. **River reversals** also occurred. Figure 4 shows how the pre-Pleistocene Ruzizi river probably joined the Rutshuru river on the Nile river drainage via Lake Edward. After rifting, the Ruzizi reversed flow, cut through an obstruction (which became the **Panzi Falls**) and then flowed into **Lake Tanganyika** from whence it joined the **Congo** river drainage which exits into the Atlantic.

In still other cases, rivers were blocked by either uplifting blocks or volcanic flows, creating lava blocked-lakes. **Lake Kivu** was blocked by volcanic flows, as were **Lake**

Bulera and **Lake Ruhondo** in Rwanda and **Lake Bunyonyi** and **Lake Mutanda** in Uganda. One of the results has been the creation of "drowned" **Ria coasts** (**Figure 4**). Other **impoundments** occurred because of tilting and warping of the Earth during the Pleistocene; e.g. **Lake Kyoga** and much of **Lake Victoria** show the results of "drowning" of river valleys. See another example in **Virtual Tour IV** online which gives a profile diagram of the **Mukungwa watershed**.

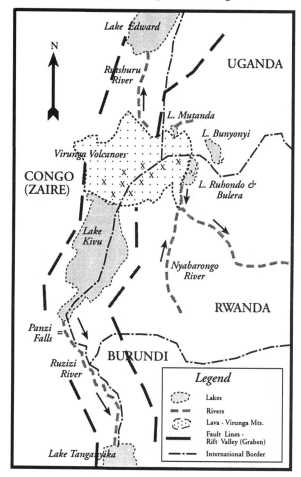

Volcanism in East Africa

Active volcanism is seen today in only three areas of East Africa: the **Virunga Mountains** of Rwanda, Uganda and the Congo; the **Afars Depression** of the **Ethiopian rift valley** (ERV); and a single volcano, named **Ol Doinyo Lengai,** on the floor of the Gregory rift valley south of **Lake Natron** in Tanzania. In fact, Eastern Africa has some of the most diverse and fascinating explosive and non-explosive volcanoes anywhere on Earth. One can also observe all types of volcanic and **pyroclastic** materials.

Figure 4. Map of the major tectonic, volcanic and hydrologic features of the Western Rift Valley (WRV) region from northern Lake Tanganyika to Lake Edward. You can see major river derangements, lava-blocked lakes, volcanic lava fields, and the lineaments (escarpments and grabens) associated with the Western Rift Valley.

How are vegetation, climate and soils influenced by the underlying geomorphology, geology, and hydrology?

The distribution of **soils** and **vegetation** on a landscape is the most visible portion of the **biosphere**—the "life layer". The quality, fertility and productivity of these **biotic** and **abiotic** systems largely define the **ecosystems** and **habitats** that are found there. The patterns of interaction between organisms in these systems—**ecosystem dynamics**—in turn, directly impact natural resource availability as well as ecosystem fragility, stability, productivity and resilience.

Soil and vegetation patterns, and their associated animal life, also reflect many factors: local and regional patterns of climate (temperature, precipitation), parent rock from which the soils form, relief and slope characteristics of the landforms, altitudinal/latitudinal situation and location, and aspect (orientation toward the sun or rain-bearing winds). All these factors are related to surface and sub-surface water patterns—the types of streams and lakes observed, whether they are perennial or ephemeral, how plentiful and accessible groundwater is, and so on. In the Virunga region of Rwanda, and elsewhere along the East African rift system, we can see many examples of this tight link between the landscape, ecology, and natural resources.

Climate and Weather in East Africa

In East Africa, regional and local climate patterns are influenced by their position in relation to major air masses, land-water interactions, the seasonal movement of the ITCZ (Inter-Tropical-Convergence Zone), latitude and elevation above sea level. On a regional basis, this produces an alternating pattern of wet and dry seasons—called the *Big and Little Rains*—and it means a generally increasing amount of moisture and decreasing temperature with altitude. Local relief factors also create unique patterns of rainfall and vegetation, such as the local rain-maker effect of lakes and mountains, or the barrier-effect of mountains and escarpments that create rainshadows.

Soil and Water

Soil fertility and their susceptibility to "degradation" from erosion is affected by factors such as slope, parent rock sources, physical and chemical weathering, ground and surface water patterns, climate patterns (temperature, precipitation) as well as the intensity and type of human use of the land, e.g. agricultural patterns, fire, hunting, forestry, fishing. Access to surface water, whether in lakes or streams or to groundwater through wells is also heavily impacted by a complexity of factors. In areas with recent volcanism, such as the Virunga mountains, surface and groundwater is very intermittent or difficult to access due to the porous nature of the lava fields. Groundwater is almost non-existent. On the other hand, the lava soils are among the richest on the continent; recall that for most of Africa soils are very old, eroded, and highly weathered, which generally means low fertility.

Flora and Fauna

The **flora and fauna** of these ecosystems reflect the above factors and more. Of particular concern in much of East Africa is the rapid disappearance of the highly diverse and often unusual plant and animal species, refered to as biodiversity loss. Some of the species seen on high volcanic peaks such as **Kilimanjaro** in Kenya/Tanzania and **Karisimbi** in the **Virunga** mountains as well as **Mount Kenya** and the **Ruwenzori** mountains are "living fossils," or species which have survived major climate and tectonic changes dating back to the Pleistocene. No wonder these rare volcanic peaks are called "**islands in the sky**" by biologists.

The plight of big-game wildlife in East Africa has become a global conservation problem, whether it be the mountain gorilla in the Virungas or the white rhino and African elephant. In mountain regions like the Virungas, plants and animals organize themselves in vertical communities and niches that relate closely to specific altitudinal zones that reflect variations in temperature, moisture, insolation and wind exposure. This adds to the beauty and uniqueness of the landscapes.

How have humans interacted with this landscape and what are the implications for sustainable development?

The Virunga mountains and the surrounding **Great Lakes Region** have some of the most densely settled rural areas of the African continent outside the Nile Delta. Some of these countries are experiencing extremely high population growth. As a consequence, one of the **key questions** driving earth system science research in East Africa today is whether current human-environment interaction patterns—particularly "population-environment relations"—is out-of-balance with the underlying natural resource base. In other words, are current patterns of exploitation and use of resources leading toward **sustainable development** or not?

Another even more crucial question is: who is most vulnerable to environmental degradation and the human insecurity that leads to environmental scarcity or even natural or human-induced disasters? Considerable empirical and applied research suggests that vulnerability to human-induced or natural system risk is not equally distributed geographically or by social group. Getting answers to these questions is very difficult but is absolutely crucial to human survival in the region.

Of course, many complex and interrelated factors are involved and we must be careful to avoid jumping to simplistic linear conclusions when linking human population variables to the environment. Yet, one cannot deny that geology has been a primary factor in producing the diverse natural resource base upon which society is based. The rich volcanic soils, the varied relief and landforms, the fantastic spatial distribution of micro-climates and ecosystems can all be traced back to its volcanic and tectonic history. Many factors have favored the evolution of high animal and plant species diversity in the region: the zone's location close to the Equator, the year-round exposure to sunlight and the minimal variation in day-length favors almost year-around plant growth. Some "high potential" areas here in East Africa can produce three different agricultural crops a year.

There are many causal factors which have contributed to producing the right conditions for the high population density and agricultural productivity of this region, particularly in what some have called "high potential" areas. The diverse ecological and physical forces in the region have also produced less favored "low potential" areas, to use terms that date back to colonial times. What is fascinating is how "low" vs. "high" potential areas are often in close proximity. This means that maps of land use and land cover in East Africa are complex and show dramatic spatial variability over short distances.

As to human comfort and health, as well as attractiveness for tourism—the middle to high mountain altitudinal zones are without question among the most ideal climates anywhere on Earth. Many favored areas are above the disease-laden lowlands where vectors such as the mosquito or *Tsetse* fly transmit diseases that plague humans and animals. Unfortunately, it appears that rapid human and ecological changes occurring in the region may be related to increasing resistance by the disease carriers to modern drugs and medicines. Not surprisingly, therefore, the highest population densities in East Africa are most frequently found in the favored mid-slope zones.

There are many complex cultural, historical, political and economic factors involved in explaining the high population density and landscape ecology story in East Africa. Not everything can be explained by environmental factors alone. And, whether the high population density and growth is a **problem or opportunity** is the object of considerable debate among academics and policy-makers alike. Yet, all would agree that underlying the complex "human" factors is the unmistakable imprint of the geo-ecological forces upon which the **nature-society relationships** in the region are based.

How is human vulnerability and risk (such as land degradation) monitored and mitigated using "earth systems science" tools and techniques such as GIS, remote sensing, and global positioning systems?

One of the most critical problems in doing GEC (Global Environmental Change) research is the lack of good spatial and temporal information about climate, soil and vegetation patterns, animal distributions and migrations, and of course, human impacts on land use and land cover. The physical and human complexities force us to look for tools that are cheaper, more timely, effective and much more extensive in their reach. Hence the need and opportunities afforded by the use of modern "hi-tech" tools such as GIS (geographical information systems) and GPS (global positioning systems) to map, monitor, manage and mitigate the "driving forces" of "global change" that are accelerating land degradation, biodiversity loss, erosion, deforestation, land use and land cover change (LULC) in the region. One of the most fascinating examples of this new "hi-tech" approach is the use of GPS by wildlife managers and scientists in the Virunga National Park to monitor mountain gorilla migration as well as the location of poachers.

Other research projects you might want to explore which focus on the "human-environment interaction" issues of the region include: the several publications of the **CIESIN/Rwanda Society-Environment Project** that was carried out by researchers at Michigan State University from the late 1980s until mid-1990s. The work by Michael Mortimore and Mary Tiffen on the Machakos District of Kenya has spawned considerable debate. Their analysis has provided a more positive counterpoint to the often "pessimistic" view of population trends in the region. This will afford you a very brief but fascinating introduction to some very exciting new technologies as well as theories and methodologies within earth system science and related disciplines.

CONCEPTUAL QUESTIONS

What are the principal natural resource endowments and physical features of the East African landscape that can be traced back to its complex volcanic and tectonic history? Can you explain the processes that contributed to that complex history?

What do *you* think are the answers to some of the troubling resource management and ecological issues raised in the case that relate to population-environment interaction as well as HDGC — the human dimensions of global change?

Is the achievement of sustainability a viable and achievable scenario for the countries and peoples who live along the Western Rift Valley, particularly in the Great Lakes Region? Or are these peoples doomed to a "slippery slope" that will spiral out of control bringing more conflict, land degradation, overpopulation and death?

How would an uninformed scientist or policy-maker who wants to be both objective but also concerned and involved go about seeking the answers?

SOME RECOMMENDED READINGS

Books:

Beadle, L.C. 1981. *The Inland Waters of Tropical Africa: An Introduction to Tropical Limnology*. (2nd Edition) Longman.

de Blij, Harm J. 1997. *Geography: Realms, Regions, and Concepts*. John Wiley & Sons, Inc.

Ford, Robert E. 1997. Settlement Structure and Landscape Ecology in Humid Tropical Montane Rwanda. In *Rural Settlement Structure and African Development*. Edited by Marilyn Silberfein. Boulder: Westview Press.

Ford, Robert E. 1993. Marginal coping in extreme land pressures: Ruhengeri, Rwanda. In *Population Growth and Agricultural Change in Sub-Saharan Africa*. Edited by B.L. Turner II, Robert Kates, and Goran Hyden.University Press of Florida.

Gregory, J.W. 1896 (1968). *The Great Rift Valley*. Republished by Frank Cass.

Nyamweru, Celia. 1980. *Rifts and Volcanoes: A Study of the East African Rift System*. Thomas Nelson and Sons.

Simkin, T. and L. Siebert. 1994. *Volcanoes of the World*. Tucson, Arizona: Geoscience Press.

Skinner, Brian J. and Stephen C. Porter. 1995. *The Dynamic Earth: an Introduction to Physical Geology* (3rd edition). New York: John Wiley & Sons, Inc.

Skinner, Brian J. and Stephen C. Porter. 1995. *The Blue Planet: an Introduction to Earth System Science.*). New York: John Wiley & Sons, Inc.

Skinner, Brian J. and Barbara Murck. 1998. *Geology Today: Understanding Planet Earth*. New York: John Wiley & Sons, Inc.

Summerfield, Michael A. 1985. Plate tectonics and landscape development on the African continent. In: M. Morisawa and J.T. Hack (eds) *Tectonic Geomorphology*. Allen and Unwin, Boston & London, 27-51.

Summerfield, Michael A. 1991. *Global Geomorphology*. Longman Scientific.

Wilcock, Colin. 1974. *Africa's Rift Valley*. Time-Life Books.

Articles:

Bonatti, E. 1987. The rifting of continents.

Scientific American. 256(3), 74-81.

De Mulder, M. 1986. K-Ar geochronology of the Karisimbi Volcano (Virunga, Rwanda-Zaire). *Journal of African Earth Sciences,* 575-579.

Ebinger, C.J. 1989. Tectonic development of the western branch of the East African Rift System. *GSA Bulletin* 101(7): 885-903.

Ford, Robert E. 1990. The dynamics of human-environment interactions in the tropical montane agrosystems of Rwanda. *Mountain Research and Development.* 10 (1): 43-63.

Hunger Notes (Summer 1996) - Special Issue. Rwanda: What have Humanitarians Learned? Vol. 22, No. (Special Editor, Peter Uvin, The Brown Hunger Program). World Hunger Education Service, P.O. Box 29056, Washington, DC 20017 Tel. 202-269-6322.

Lewis, Lawrence. 1992. Terracing and Accelerated Soil Loss on Rwandan Steeplands: a Preliminary Investigation of the Implications of Human Activities Affecting Soil Movement. *Land Degradation and Rehabilitation.* 3, 241-46.

MacKay, M.E. and Peter J. Mouginis-Mark. 1997. The effect of varying acquisition parameters of the interpretation of SIR-C rada data: the Virunga Volcanic Chain. *Remote Sensing of the Environment.* 59: 321-336.

Olson, Jennifer M. 1994. Demographic Responses to Resource Constraints in Rwanda. Rwanda *Society-Environment Project, Working Paper 7*, September . CIESIN / Michigan State University.

Percival, Valerie and Thomas Homer-Dixon. 1995. *Environmental Scarcity and Violent Conflict: The Case of Rwanda.* The Project on Environment, Population and Security, University College of Toronto / American Association for the Advancement of Science.

Ruff, Nathan and Robert Chamberlain and Alkexandra Cousteau. 1996. Report on Applying Military and Security Assets to Environmental Problems. *Environmental Change and Security Project Report - The Woodrow Wilson Center.* Spring 1997 Issue 3: 82-95.

Tiffen, Mary and Michael Mortimore. 1994. Environment, population growth and productivity in Kenya: a case study of Machakos District. *Drylands Network Programme: Issues Paper No. 47. January.*

Uvin, Peter. 1995. *Genocide in Rwanda: The Political Ecology of Conflict.* Research Report / The Alan Shawn Feinsteing World Hunger Program, Brown University. RR-95-2.

White, R.S. and McKenzie, D.P. 1989. Volcanism at rifts. *Scientific American* 261,1:44-55

Case 8

How Much Topsoil is Left in the Corn Belt? Land and Life on the North American Prairie

Questions Explored:

What is the grassland biome and where is it located around the world and in North America?

How has the interaction of geologic, hydrologic, climatic, biotic, and human forces through time produced the Mid-Continent Prairie of North America and its associated soils and biota?

How have humans impacted the prairie landscape through time and particularly the subregion known as the Corn Belt?

How is land degradation—particularly soil erosion, groundwater extraction and biodiversity loss—monitored and mitigated using "earth systems science" (ESS) tools and techniques

Key Learning Outcomes:

The learner will be able to:

Locate on a world map the major subdivisions of the grassland biome including the tall and short-grass prairies of Mid-Continent North America.

Locate on a map of North America the major physical geographical, cultural and agricultural subregions of the Great Plains/Interior Lowlands.

Identify from maps, photos and remotely sensed imagery some of the geological, soil, vegetation, hydrologic, glacial, settlement, agricultural and rangeland patterns that typify the landscapes of Mid-Continent of North America.

Discuss and evaluate scientific findings regarding sustainable development and natural resource conservation issues, particularly soil erosion and biodiversity loss, affecting the Mid-Continent Prairie.

OVERVIEW

Location

In this learning activity you will explore the **grassland biome** of the world and particularly that portion of **midlatitude North America** known as the **Great Plains** or **Mid-Continent Prairie**. There are other important midlatitude, subtropical, and tropical grasslands each with their own distinct history, ecology and local names. These include, for example the **Pampas** in South America, the **Veld** in South Africa, the **Chernozems** of Ukraine and Belorus and the **savannas** and **steppes** of the arid and semi-arid tropics.

Author:

Robert E. Ford
Westminster College of Salt Lake City

Primary Objectives

The primary objectives of this case study are to:

a) introduce students to the underlying biophysical forces which characterize the global distribution of the **grassland biome** including the western **rangelands** and **prairies** of North America,

b) consider in more detail how the **Prairie/Great Plains** ecoregion of Mid-Continent North America has evolved through time, particularly during the **Holocene**, and,

c) explore some of the complex issues of **land degradation** and **ecosystem management**—particularly **soil erosion, groundwater extraction** and **biodiversity loss** affecting the region.

The **key question** to be considered is the following: **have the limits to sustainable development been breached?** If so, how can we measure, monitor and mitigate the forces which produce **land degradation** and particularly **topsoil** and **biodiversity loss** in the world's grasslands and rangelands and can we **restore these ecosystems?**

BACKGROUND

What is the grassland biome and where is it located around the world and in North America?

In simple terms, the **grassland biome** includes a broad collection of **ecosystems** that exist between the drier margins of the forest margin and the edge of the desert. Grasslands, like other biomes, share certain common ecological characteristics due primarily to adaptation to soil, water, climate, fire and other constraints on plant growth.

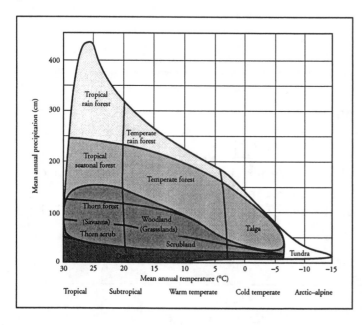

Figure 1. World biomes in relation to altitude, latitude, precipitation and temperature conditions.

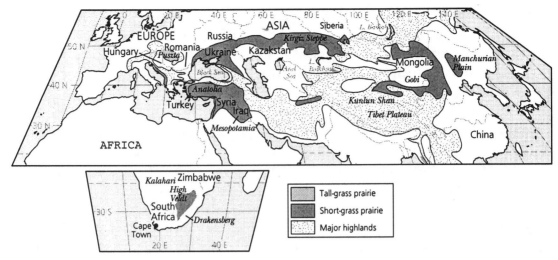

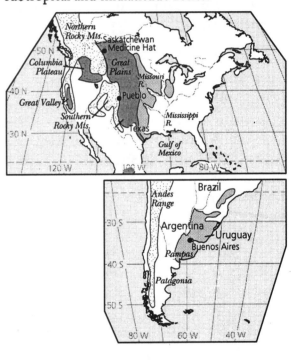

Precipitation effectiveness, a critical indicator of whether a given area will support a forest, grassland, desert or tundra, is most frequently estimated by measuring **evapotranspiration** and computing an **annual water budget** for a given area. **Temperature** and seasonal **solar relationships**; i.e. orientation toward the sun is also important because both temperature and precipitation conditions together determine general plant growth conditions. Therefore, conditions that favor grasslands over forest also vary as one changes **elevation** or where **fire** is more frequent. **Figure 2** shows the distribution of the grassland biome in the sub-tropics and mid-latitude zones. Note that the grasslands are located in transition zones between the forest boundary and the arid lands.

Figure 2. World map of the grassland biome in subtropical and midlatitude zones.

How has the interaction of geologic, hydrologic, climatic, biotic, and human forces through time produced the Mid-Continent Prairie of North America and its associated soils and biota?

The Grassland Ecosystem

The **Great Plains Grasslands Ecoregion** in North America, also known as the **Mid-Continent Prairie,** illustrates well the tight link that exists between landscape evolution, geology, ecology, climate and natural resource availability and productivity which produces the ecosystems of a biome. While some observed features and processes reflect shorter or longer term natural fluctuations in the biophysical environment resulting from climate change, other processes can be attributed to human impact such as land use and land cover change or economic activity under varying technological conditions.

Human use of **fire**, for instance, dating back to prehistory has greatly expanded the grasslands. Later, **grazing** by domesticated animals further altered the grasslands. In fact, in both the tropical **savanna grasslands** and in the **midlatitude steppes** and **prairies, anthropic influences** on both the structure and species composition of the grasslands has been very significant.

Grasslands around the world are not alike though grasslands in like environments often share similar physiognomic characteristics. Their appearance reflects constant adaptation of plant formations and individual species to drought, fire, animal predation, climate change and human impact. Grasslands also exhibit great species diversity and support some of the most complex food webs of any ecosystem on Earth. Picture, for instance, the tremendous animal species diversity and awe-inspiring competition for resources along the trophic chain in the savannas of the Serengeti Plain in Tanzania. This gives just an inkling of what the North American Great Plains environment must have been like during wetter phases of glacial times when the Mastadon and Sabre-Tooth Tiger roamed among millions of Bison and other now extinct creatures.

Botanically the grasslands biome is dominated by grasses but the story is actually much more complex. Undisturbed natural grasslands include many **non-grass species,** including **forbs** as well as **woody shrubs**, and trees. The structure of grassland **vegetation formations** is also very complex. There is complexity in how species are organized spatially varying from north to south as well as from east to west. In North America, one can observe a whole range from short-grass **steppes, scrublands** and **prairies** on the dry arid boundary to tall-grass prairies, **savannas, woodlands** and **forests** on the humid boundary.

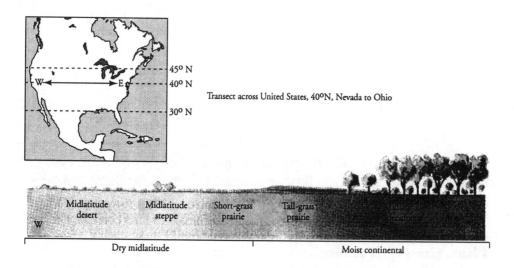

Figure 3. A continental transect across the prairie region of North America showing the succession of plant formation across climatic gradients.

Grasslands or prairies rarely exist in continuous stretches. More often they are organized as a **mosaic** of patches interspersed between **gallery forests** along streams or among other plant formations depending on local soil-water-climate conditions. This patchiness is particularly pronounced where there is greater **local relief**, where the underlying geology is complex, where fire has not been controlled, along the forest or desert boundary zone, and where unique climatic conditions related to glacial history have created locally unusual **edaphic** or soil-water conditions. Examples of the latter are the **sloughs** of North Dakota or **bogs** of Wisconsin and Minnesota. Many are **kettles and kames,** a type of glacial feature common to **outwash till plains.** These **wetlands** are extremely important breeding grounds for **water birds** such as ducks and geese and they frequently feed on plants growing in adjacent grasslands, scrub or woodlands.

In sum, the grassland ecosystem is a much more diverse landscape than many expect. The prairie, in fact, has a unique beauty and appeal that has inspired many great natural history writers and even novelists and historians, including Aldo Leopold (1949) *A Sand County Almanac (* New York: Oxford University Press), Diane Quantic (1995) *The Nature of the Place: a Study of the Great Plains Fiction* (Lincoln, NE: University of Nebraska Press), and Wilbur R. Jacobs (1994) *On Turner's Trail: 100 Years of Writing Western History* (Lawrence, Kansas: University Press of Kansas).

Earth History and Landscapes on the Great Plains
Local relief on the Great Plains is quite low, nevertheless, earth forces have produced a much more varied landscape than most expect. In places such as Minnesota, Illinois, the Dakotas and Wisconsin much of the High Plains were once covered partially or completely by **continental glaciers,** in some cases several times. Landscapes more distant from the glacier margins were affected by **eolian** (wind) processes; e.g. creation of **loess deposits** and **sand dunes** such as along the South Platte River in Nebraska. Many of the region's richest farming soils are on these wind-blown silty soils.

Farther south lowland **river basins** and extensive **floodplains** experienced intense **accelerated erosion** or **alluvial deposition** by the swollen streams and rivers flowing from the **alpine glaciers** that covered the **Rocky Mountains** or from the **continental ice sheets** themselves. Even periods of **invasion by the rising sea** were important for certain places such as in the lower Mississippi Valley. To a lesser degree **tectonic mountain building** and **volcanism** have influenced some landscapes, though much more occurred in earlier geological periods.

The glacial and post-glacial period of earth history has produced a wide range of landforms that significantly altered the underlying geology and thus the distribution of soils, vegetation and animals. On the website accompanying this case study are many photos, maps and images that show some of the diversity in landscapes and landforms of the Great Plains. These include for example: **moraine** and **eskers** in areas along the glacier margins as well as **till plains, kames and kettles** in the **peri-glacial** zones (**outwash plains**) as well as areas that experienced no glaciation, such as the **Driftless Area.** The **Nebraska Sand Hills** is also a remnant feature that owes its existence to this complex earth history.

Environmental Change on the Prairies

As the glaciers retreated during the Holocene, and the climate became more arid, the Great Plains offered many opportunities for plant and animal **migration, diffusion** and **extinction**. Further on in time humans introduced new animal and plant species or killed off others. And as **rivers were diverted, wetlands drained,** and **groundwater extracted** the process of environmental change accelerated. Some of the introduced plants have become veritable weeds and have driven out or replaced many of the original prairie grasses and forbs. Thankfully, many dedicated volunteers across the Great Plains are today trying to rediscover and save many of the "native" prairie plants.

A critical climatic precipitation boundary on the Great Plains roughly parallels the 100th Meridian. Throughout history this **critical arid/humid boundary** has shifted back and forth—the result has frequently been disastrous. The **Dust Bowl** experience was a severe lesson to many early pioneer farmers who did not understand the vagaries of "normal" climate cycles on the dry plains.

THE HUMAN DIMENSION

How have humans impacted the prairie landscape through time particularly in Agriculture and the Corn Belt?

Agriculture and the Corn Belt

The North American **tall-grass prairie** and particularly the subregion known as the **Corn Belt** is by far the world's greatest producer of **cereals** such as maize and sorghum (milo). Some of the cereals are eaten directly by humans but much is fed to beef cattle and pigs. The short-grass prairie also produces vast amounts of spring and winter wheat. And on the drier margins of the High Plains and in the western uplands and mountains cattle and sheep are still grazed on the "open range" or rangelands.

Today, human impacts in the Corn Belt, resulting from the attempted management of entire drainage basins or ecosystems and the massive mining of topsoil and groundwater resources by industrial agriculture and animal husbandry rivals if not surpasses what natural forces or Early Man did in the past. A few small "restored" prairies and National Grasslands remain but nothing rivals the vast sea of grass that existed before the sod-busters and cattlemen arrived. (You are encouraged to explore the web-based material produced by the National Homestead Monument and other sites such as the Illinois Science Museum which depicts the pioneer history of the Great Plains. See also Bogue, Allan G. 1994. *From Prairie to Corn Belt: Farming on the Illinois and Iowa Prairies in the 19th Century*. Ames, IA.: Iowa State University Press; Jacobs, Wilbur R. 1994. *On Turner's Trail: 100 Years of Writing Western History* (Lawrence, Kansas: University Press of Kansas).

But the problem is even more complex than just managing basic soil and water resources for farmers and ranchers—the prairies and rangelands are not just for producing beef and pork. Prairies are increasingly important recreational resources as well that provide access to open space and wilderness for people living in urban regions. And they are critical reserves for preserving the **genetic biodiversity** that is essential

84

for modern industrial agriculture and medicine.

On both the High Plains and in adjacent mountain lands, management of "non-traditional" natural resources such as open space, watersheds and wilderness increasingly competes with older extractive industries such as ranching, mining, and agricultural production. This has increased the level of conflict between competing interest groups. Solving and managing these complex issues (ecosystem management) requires a broad perspective by many specialists working together in teams and using multiple tools and techniques. Learning to do this type of integrative and interdisciplinary study is what Earth Systems Science (ESS) is all about.

The critical questions that remain are the following: have humans so altered these grassland ecosystems that they can no longer be restored? Is the unique soil-forming process that created the deep layers of dark topsoil no longer in balance? And are we capable of managing what is left in a sustainable manner into the future?

How is land degradation, particularly soil erosion and biodiversity loss, monitored and mitigated using "earth systems science" (ESS) tools and techniques?

Land Degradation and Soil Erosion

Humans have interacted with grasslands in very complex ways since the beginning of time. Each period of **human occupance** left an indelible imprint on the landscape. In some cases human impacts have been minimal and sustainable while other interventions have completely altered or even replaced the original ecosystem. A particularly severe type of **land degradation** which some people call **desertification** is a problem of global proportions. Desertification is an insidious and often irreversible process of deterioration of soils, water, vegetation and wildlife resources which is complex from both a biophysical and human perspective.

One of the most critical conservation issues facing the grasslands is how we manage and use the topsoil found in the Corn Belt. To better understand this issue a brief review of the basics of soil science or pedology and soil classification as found in any basic physical geography, ecology or geology textbook is recommended. First, review the famous formula by Hans Jenny (the father of modern soil science) that outlines the critical soil-forming process or soil factors:

$$s = f(cl, o, r, p, t, \ldots)$$

Recall that the s is for soil property, cl for climate, o for organisms, r for relief, p for parent material and t for time. In simplest terms a given soil is a function, f, of these five soil-forming factors. This complex process creates a mosaic of soil types that vary dramatically over the landscape. In addition, one needs to understand that some soil-water-vegetation combinations are more fragile than others. To assess this, land managers create special land capability maps as well as basic soil series maps that combine many factors to assist in both managing, and preserving the land base or restoring abused land.

At one time the Corn Belt boasted a deep brown to black layer of decomposed grasses and roots (the humus layer) that in some places was 20-30 feet deep. According to some estimates, only one half of the original topsoil layer remains only a century after the tall-grass prairie was put to the plow. What has caused this radical topsoil loss? Soil scientists call this process accelerated soil erosion, that is, erosion in excess of the "normal" balance between soil loss and soil formation that goes on all the time. This type of soil erosion is clearly a case of soil mining for short term productive gain and is a practice that is not sustainable!

Earth scientists who study soil erosion, either by wind or water, know that erosion rates vary greatly from place to place. They usually differentiate between erosivity of raindrops and running water on the one hand, and the erodibility (i.e. detachability and transportability) of soil material on the other. Much of soil erosion research is concerned with measuring and comparing the variables that determine these forces in order to be able to predict the likelihood of erosion and reduce soil loss on both agricultural as well as range or forested landscapes.

The spectre of declining agricultural productivity arising from erosion of soil by wind and water has haunted conservationists for much of the twentieth century. Intimations of impending catastrophe and appeals for action have been broadcast for many years through such books as *The Rape of the Earth* by Jacks and Whyte (1939), *Bennett's Soil Conservation* (1939) and the FAO's *Soil Erosion by Water* (1965). The plight of the "**Okies**" and others displaced by the **Dust Bowl** is still engraved in most people's minds as *the* most visible ecological disaster in living memory. The very creation of what was formerly known as the Soil Conservation Service (now the Natural Resource Conservation Service (NRCS)) dates back to this major human-induced disaster.

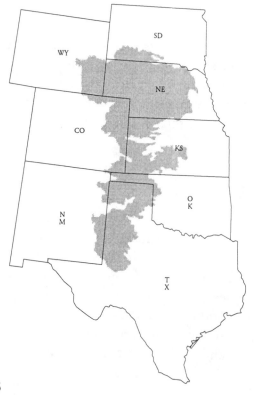

The lowering of the **Ogallala or High Plains aquifer** because of over-pumping for agriculture, though not as visible, may be producing as equally serious a disaster as was the Dust Bowl (Figure 4). Large scale mining of the fossil water from that aquifer is depleting and polluting this non-renewable groundwater resource which dates back to the wetter phases of the Pleistocene. Furthermore, the soil itself is being altered from its previous "rainfed" character. Some authorities predict that when the water runs out these short-grass prairie soils will have been so altered that return to the previous ecological balance

Figure 4. Location of the Ogallala or High Plains aquifer.

may be impossible. Of course, much of the previous animal and plant life adapted to the prairie has been crowded out by exotic plants.

In summary, land degradation on the prairies not only threatens our **food security**, but may in fact be altering regional and global climates and ecosystems. Much of the loss is due to **soil erosion** as well as **groundwater extraction** and **biodiversity loss**. It is therefore critical that we better understand how we impact the grasslands, and more important, how we can develop less intrusive and more sustainable ways of interacting with them.

Monitoring and Mitigating Land Degradation

One of the most critical problems in monitoring and mitigating land degradation, a primary theme in **GEC** (Global Environmental Change) research, is the lack of high quality **spatial** and **temporal** information about climate, soil and vegetation patterns, animal distributions and migrations, groundwater and of course, human impacts on **land use cover change** (LUCC). The physical and human complexities of these problems, and their seriousness, force us to look for tools that are cheaper, more timely, effective and much more extensive in their reach. Hence, the need and opportunities afforded by the use of modern "hi-tech" tools such as **GIS** (geographical information systems) and **GPS** (global positioning systems) to **map, monitor, manage** and **mitigate** the **driving forces** of **global change** that are accelerating land degradation, biodiversity loss, erosion, deforestation, and **land use cover change** in the region.

CONCEPTUAL QUESTIONS

What are the principal natural resource endowments and physical features of the Great Plains/Corn Belt/Tall-grass Prairie landscape that can be traced back to its complex geologic and Pleistocene history? Can you explain the processes that contributed to that complex history?

What do *you* think are the answers to some of the troubling resource management and ecological issues raised in the case that relate to human-environment interaction as well as the human dimensions of global change?

Is the achievement of "sustainability" a viable and doable scenario for the region and peoples who live there? What is the potential of alternative methods of farming as well as urban-oriented uses of the land that will restore a balance between soil-loss and soil-formation as well as biodiversity loss?

Can land that has been abused be truly restored? What are some of the potential costs and barriers that impede ecosystem restoration activities in the region?

SOME RECOMMENDED READINGS

Books:

de Blij, Harm J. 1997. *Geography: Realms, Regions, and Concepts.* John Wiley & Sons, Inc.

Bogue, Allan G. 1994. *From Prairie to Corn Belt: Farming on the Illinois and Iowa Prairies in the 19th Century.* Ames, IA.: Iowa State University Press.

Dawson, Alastair G. 1992. *Ice Age Earth: Late Quaternary Geology and Climate.* London and New York: Routledge.

Fehrenbacher, J.B. et al. 1984. *Soils of Illinois, Bulletin 778.* Champaign: University of Illinois, College of Agriculture.

Follett, R.F. and B.A. Stewart (eds). 1985. *Soil Erosion and Crop Productivity.* Madison, WI: American Society of Agronomy.

Glanz, James. 1995. *Saving our Soil: Solutions for Sustaining Earth's Vital Resource.* Boulder, CO: Johnson Books.

Houle, Marcy Cottrell. 1995. *The Prairie Keepers: Secrets of the Grasslands.* Reading, MA: Addison-Wesley.

Hunt, Charles B. 1972. *Geology of Soils: Their Evolution, Classification and Uses.* San Francisco: W.H. Freeman and Company.

Jacobs, Wilbur R. 1994. *On Turner's Trail: 100 Years of Writing Western History.* Lawrence, Kansas: University Press of Kansas.

Leopold, Aldo. 1949. *A Sand County Almanac.* New York: Oxford University Press.

Manning, Richard. 1995. *Grassland: The History, Biology, Politics, and Promise of the American Prairie.* New York: Viking Press.

Quantic, Diane Dufua. 1995. *The Nature of the Place: a Study of the Great Plains Fiction.* Lincoln, NE: University of Nebraska Press.

Riley, Glenda. 1988. *The Female Frontier: a Comparative View of Women on the Prairie and the Plains.* Lawrence, KS: University Press of Kansas.

Samson, Fred B. 1996. *Prairie Conservation: Preserving America's most Endangered Ecosystem.* Washington, D.C.: Island Press.

Skinner, Brian J. and Stephen C. Porter. 1995. *The Dynamic Earth: an Introduction to Physical Geology* (3rd edition). New York: John Wiley & Sons, Inc.

Skinner, Brian J. and Stephen C. Porter. 1995. *The Blue Planet: an Introduction to Earth System Science.* New York: John Wiley & Sons, Inc.

Skinner, Brian J. and Barbara Murck. 1998. *Geology Today: Understanding Planet Earth.* New York: John Wiley & Sons, Inc.

Strahler, Alan and Arthur Strahler. 1994. *Introducing Physical Geography.* New York: John Wiley & Sons, Inc.

Ulrich, Hugh. 1989. *Losing Ground: Agricultural Policy and the Decline of the American Farm.* Chicago: Chicago Review Press.

USDA (U.S. Department of Agriculture). 1993. *Predicting Soil Erosion by Water. A Guide to Conservation Planning with the Revised Universal Soil Loss Equation (RUSLE).* Washington, D.C. U.S. Government Printing Office (draft) Agricultural Handbook No. 703.

Webb, W. P. 1931. *The Great Plains.* New York: Ginn.

Wilkinson, Charles F. 1992. *Crossing the Next Meridian: Land, Water, and the Future of the West.* Washington, D.C.: Island Press.

Articles:

Brown, D.A. and P.J. Gersmehl. 1985. Migration models for grasses in the American Midcontinent. *Annals of the Association of American Geographers,* 65:223-241.

Gersmehl, Philip J. 1992. The Hosmer Silt Loam in Montgomery County, Illinois, and American Farm Policy. IN *Geographical Snapshots of North America.* Edited by Donald G. Janelle. New York: Guilford Press. Pp. 69-73.

Lyles, Leon. 1995. Predicting and controlling wind erosion. *Agricultural History.* 59 (2): 205-214.

Oliver, John E. 1992. The "Great American Desert" and the Arid/Humid Boundary. IN *Geographical Snapshots of North America.* Edited by Donald G. Janelle. New York: The Guilford Press. Pp. 208-210.

Olmstead, Clarence W. 1992. A Place and Point of View: Southwestern Wisconsin. IN *Geographical Snapshots of North America.* Edited by Donald G. Janelle. New York: The Guilford Press. Pp. 337-341.

Winkler, Marjorie Green. 1992. Since the Retreat of the Ice Sheet: a Wisconsin Wetland. IN *Geographical Snapshots of North America.* Edited by Donald G. Janelle. New York: The Guilford Press. Pp. 197-203.

Image and Figure Credits

Case 1: Earthquake Hazards - The Wasatch Fault

Michael W. Hernandez
University of Utah

p. 9, Figure 1. The Intermountain Seismic Belt showing the Wasatch Fault. Note the location of Yellowstone Park, which is situated above an active volcanic hot spot.
Source: modified by Michael W. Hernandez from Arabasz, W.J., J.C. Pechmann, and E.D. Brown. 1992. Observational seismology and the evaluation of earthquake hazards and risk in the Wasatch Front area, Utah. In Gori, P.L., and W.W. Hays, eds. *Earthquake Studies in Utah.* U.S. Geological Survey Professional Paper 1500-D.

p. 9, Figure 2. Typical East-West cross section of a portion of the Basin and Range province. Stippled areas are valley-fill deposits and arrows indicate relative movement of block faults
Source: modified by Michael W. Hernandez from Hecker, Suzanne. 1993. *Quaternary Tectonics of Utah with Emphasis on Earthquake-Hazard Characterization.* Utah Geological Survey Bulletin 127.

p. 10, Figure 3. The Wasatch Fault divided into its 10 major segments. Segment names are the boldfaced text to the right of the arrows pointing to the segments.
Source: modified by Michael W. Hernandez from Machette, M.N., S.F. Personius and A. R. Nelson. 1992. Paleoseismology of the Wasatch fault zone—A summary of recent investigations, conclusions, and interpretations, in Gori, P.A., and W. W. Hays, eds. *Assessing regional earthquake hazards and risk along the Wasatch Front, Utah:* U.S. Geological Survey Professional Paper 1500, Chapter A, p. A1-A72.

Case 2: The 1993 Floods on the Mississippi and Missouri Rivers

W. Scott White
Weber State University

p. 20, Figure 1. The Upper Mississippi River Basin (UMRB) affected by the 1993 floods is spread over nine Midwestern and Great Plains states.
Source: drafted/designed by W. Scott White

p. 21, Figure 2. Stream hydrograph – USGS gauging station on the Mississippi River, St. Louis, Missouri.
Source: drafted/designed by W. Scott White

p. 23, Figure 3. Landsat TM images of the St. Louis, MO area. The left image is from 7/15/92 illustrating pre-flood conditions. The right image is from 8/19/93 illustrating peak-flood conditions.
Source: drafted/designed by W. Scott White

Case 3: Petroleum Geology: Persian Gulf vs. Overthrust Belt

Robert J. Moye and Sandra A. Zicus
Salt Lake City, Utah

p. 30, Figure 1. World map showing areas of major known oil and natural gas reserves with the location of the Persian Gulf and Overthrust Belt highlighted.
Source: modified by James D. Hipple from Skinner, B. and S. Porter. 1995. The Blue Planet. John Wiley & Sons, Inc., p. 461.

Case 4: Using GIS and Remote Sensing to Monitor Urban Habitat Change

James D. Hipple
University of Missouri-Columbia

p. 40, Figure 1. The Salt Lake County research study area and its location within the State of Utah.
Source: drafted/designed by James D. Hipple

p. 41, Figure 2. Percent total US population by residential category.
Source: drafted/designed by James D. Hipple

p. 42, Figure 3. Salt Lake County population trend from 1950 to 1990.
Source: derived from US Census Bureau statistics by James D. Hipple

p. 43, Figure 4. Biological zonation within the urban environment.
Source: modified by James D. Hipple from Dorney, Robert S. 1979. The ecology and management of disturbed urban land. *Landscape Architecture*. 69 (3):268-272, and Ridd, M. K., J. A. Merola, and R. A. Jaynes. 1983. Detecting agricultural to urban land use change from multispectral MSS digital data. *Proceedings of the ASP-ACSM Fall Convention*, (Salt Lake City, Utah: American Society of Photogrammetry and Remote Sensing), 473-482.

Case 5: Environmental Hazards on the US-Mexico Border: The Case of Ambos Nogales

Mark V. Finco and George F. Hepner
University of Utah

p. 50, Figure 1. The US-Mexico Border Region with a focus on Arizona and Sonora.
Source: drafted/designed by Mark V. Finco and George F. Hepner

p. 52, Figure 2. Ecoregions along the Arizona/Sonora border.
Source: drafted/designed by Mark V. Finco and George F. Hepner

p. 53, Figure 3. Above is a digital elevation model that shows the road network superimposed on a terrain model of the region.
Source: drafted/designed by Mark V. Finco and George F. Hepner

p. 54, Figure 4. A population density map of Ambos Nogales.
Source: drafted/designed by Mark V. Finco and George F. Hepner

p. 54, Figure 5. A map of land use patterns in the region with the sites of the industrial facilities shown.
Source: drafted/designed by Mark V. Finco and George F. Hepner

p. 56, Figure 6. Composite spatial distribution of human vulnerability.
Source: drafted/designed by Mark V. Finco and George F. Hepner

Case 6: Saline Lakes and Global Climate Change

Sandra A Zicus
Salt Lake City, Utah

p. 60, Figure 1. World map with the following major lakes located - 1-Mono Lake (California, USA), 2-Great Salt Lake (Utah, USA), 3-Laguna Colorada (SW Bolivia), 4-Laguna Mar Chiquita (Cordoba, Argentina), 5-Dead Sea (between Jordan & Israel), 6-Caspian Sea (boundary S.E. Europe and S.W. Asia), 7-Aral Sea (border between Kazakhstan & Uzbekistan,), 8-Lakes Natron, 9-Magadi, and 10-Elementeita (East Rift Valley - East Africa), 11-Lakes Amadeus and 12-Eyre (central Australia), and 13-Lake Vanda (Wright Valley, Antarctica).
Source: drafted/designed by James D. Hipple

p. 67, Figure 2. The above diagram is a chronology of Aral Sea changes from 1960 through the year 2000 (estimated). Since 1960 the Aral Sea has lost two thirds of its volume and half of its surface area.
Source: drafted/designed by James D. Hipple from data obtained from the United Nations Environmental Programme

Case 7: Landscape and Life along the East African Rift: the Virunga Mountains, Rwanda

Robert E. Ford
Westminster College of Salt Lake City

p. 71, Figure 1. The Afro-Arabian rift system.
Source: drafted/designed by Robert E. Ford

p. 72, Figure 2. A cross-sectional diagram of East Africa from the Kenyan coast to Congo Basin showing zones of up and down warping and faulting as well as major landform features: rift valleys, volcanoes, escarpments and lakes.
Source: drafted/designed by Robert E. Ford

p. 72, Figure 3. Diagram of half-grabens and drainage disruptions on Lake Tanganyika.
Source: modified from Summerfield, Michael A. 1991. *Global Geomorphology*. (Longman Scientific) by Robert E. Ford.

p. 73, Figure 4. Map of the major tectonic, volcanic and hydrologic features of the Western Rift Valley (WRV) region from northern Lake Tanganyika to Lake Edward. You can see major river derangements, lava-blocked lakes, volcanic lava fields, and the lineaments (escarpments and grabens) associated with the Western Rift Valley.
Source: drafted/designed by Robert E. Ford

Case 8: How Much Topsoil is Left in the Corn Belt? Land and Life on the North American Prairie
Robert E. Ford
Westminster College of Salt Lake City

p. 80, Figure 1. World biomes in relation to altitude, latitude, precipitation and temperature conditions.
Source: Botkin, D. A. and E. A. Keller. 1995. Environmental Science, 2nd Edition. John Wiley & Sons, Inc. p. 134.

p. 81, Figure 2. World map of the grassland biome in subtropical and midlatitude zones.
Source: Strahler A. and A. Strahler. 1998. Introducing Physical Geography, 2nd Edition. John Wiley & Sons, Inc. p. 226-227.

Figure 3. A continental transect across the prairie region of North America showing the succession of plant formation across climatic gradients.
Strahler A. and A. Strahler. 1998. Introducing Physical Geography, 2nd Edition. John Wiley & Sons, Inc. p. 232.

Figure 4. Location of the Ogallala or High Plains aquifer.
Source: drafted/designed by Robert E. Ford

Notes

Notes

Notes

Notes

Notes

Notes

Notes

Notes

Notes

Notes

Notes